REFLECTORS AND LOCALIZATION

Application to Sheaf Theory

F. VAN OYSTAEYEN
A. VERSCHOREN

University of Antwerp U.I.A.
Antwerp, Belgium

MARCEL DEKKER, INC. NEW YORK AND BASEL

Library of Congress Cataloging in Publication Data

Van Oystaeyen, F. [Date]
Reflectors and localization.

(Lecture notes in pure and applied mathematics ; v. 41)
Bibliography: p.
Includes index.
1. Noncommutative rings. 2. Sheaves, Theory of.
3. Functor theory. 4. Localization theory.
I. Verschoren, A., [Date] joint author. II. Title.
QA251.4.097 512'.4 78-31333
ISBN 0-8247-6844-2

MARCEL DEKKER, INC.
270 Madison Avenue, New York, New York 10016

Current Printing (last digit):
10 9 8 7 6 5 4 3 2 1

PRINTED IN THE UNITED STATES OF AMERICA

To Danielle and Linda

ACKNOWLEDGMENTS

We thank the University of Antwerp U.I.A. for facilities and support.

Nina's typing is impeccable as always; all errors are due to the authors and the printer's devil.

The second author wishes to thank the Department of Mathematics of the University of Utrecht for their generous hospitality. He also acknowledges financial support by NFWO Grant A2/35.

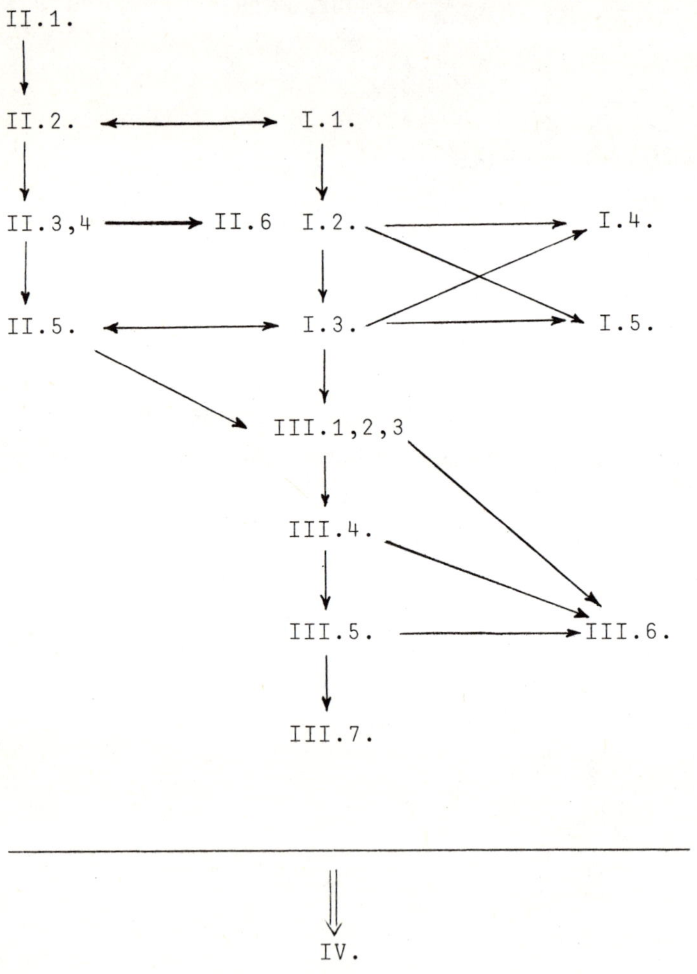
II.1.
II.2.
I.1.
II.3,4
II.6
I.2.
I.4.
II.5.
I.3.
I.5.
III.1,2,3
III.4.
III.5.
III.6.
III.7.
IV.

CONTENTS

INTRODUCTION

In Commutative Algebra, localization techniques have been the main tool in constructing sheaves over certain topological spaces and also in deriving local-global results. For a non-commutative ring R a sheaf Spec R may be constructed in such a way that many basic properties, holding in the commutative case, still hold. However, Spec is not necessarily a functor from the category of rings to the category of sheaf-spaces. One of our aims in this paper is to defend the thesis that this functoriality is not where it is at, but Spec may be regarded as a functor in a category of (pre) sheaves of modules over the fixed ground Ring, i.e. the underlying topological space together with its structure sheaf of rings. From our point of view, the deeper reason why Spec works so well in the commutative theory is that the Zariski topology allows a basis of t-sets, cf. Chapter I, what makes it possible to derive some results about ring-objects in the Grothendieck category under consideration. So, in the general case, we are led to the consideration of kernel functors and localization in categories of (pre) sheaves of modules.

Sections I.1. and I.2. introduce the elementary theory of rings, modules and localization needed in the sequel, then in Section I.3. we construct certain (pre) sheaves of rings and modules over the prime spectrum and over the torsion theoretic spectrum of a left Noetherian ring. The final two sections of the first chapter are concerned with the investigation of two special types of rings. Fully left bounded left Noetherian rings are important because localization may be fully described using ideals; moreover the prime spectrum and the torsion theoretic spectrum of such rings coincide. Zariski central rings are of interest because they constitute a class of (non-commutative) rings R for which the Zariski topology on spec R has a basis of t-sets. The special rings and the (pre) sheaves constructed over the topological spaces associated to them, which are being studied in the first chapter, serve as an example-mine for the theory expounded afterwards.

Results on localization of (pre) sheaves may be interpreted as being results on Giraud subcategories of a Grothendieck category, i.e. using the reflector associated to a Giraud subcategory, we induce kernel functors in the Giraud subcategory, starting from a kernel functor in the given Grothendieck category. In Chapter II, after an introductory section, we look at some new results relating the localization functor associated to a kernel functor in a Grothendieck category, to the localization functor associated to the induced kernel functor in some Giraud subcategory. The standard example of this is the "embedding" of the category of sheaves into the category of presheaves. The underlying fact, making possible everything we want, is that an injective sheaf is injective as a presheaf. So the formation of a Ring or Module of quotients presents no difficulties; however we have to impose restrictive conditions on the ground Ring if we want the localization of the ground Ring to be a Ring.

Two special types of kernel functors press forward. A local kernel functor K is described by giving a set of ordinary kernel functors K(U), one for each open set U in the underlying topological space X, such that they "stick together well". A strictly local kernel functor K is given by a bunch of ordinary kernel functors K_x, one for each point $x \in X$, which also stick together well.

In the local case, the localization functor Q_K associated to K is also described locally and it has the property that $Q_K(R)$ is a Ring in case R is a flabby presheaf of rings such that K reduces R. The latter condition may be dropped if only global sections are being taken into account. In Section III.6. we develop a localization technique, starting from the philosophy that global sections should be the screen where everything about localization can be read. Consequently, flabby presheaves will be the tools of main importance here. In order to get rid of the flabbiness hypothesis one has to step over to the strictly local case. This may be done if one can calculate "stalks" for the kernel functor K by $K_x = \varinjlim_{x \in U} K(U)$ or some formula like it. If the ground Ring is a coherent sheaf of left Noetherian rings, then every local kernel functor allows "stalks". Coherence and quasi-coherence and the relation of these concepts with strictly local kernel functors is being treated in Section III.7. These results actually allow many calculations to be carried out in a practical manner, in most concrete examples.

Representation of an object (ring) by the global sections of a sheaf is a well known technique. Pointwise localization as developed in Section III.6. obviously has consequences in this theory and it awaits further application.

We have paid particular interest to kernel functors K for which the corresponding localization functor Q_K is right exact and commutes with direct sums. In all cases considered we investigate conditions equivalent to this, and as it turns out, these so-called T-functors, do not only generalize Spec of a commutative ring but they behave nice in every way e.g. with respect to formation of the quotient category, quasi-coherence and coherence etc...

In the last chapter we present some applications in noncommutative algebraic geometry, introducing schemes which behave in a nice, functorial way with respect to extensions.

Our presentation is as self-contained as possible. Localization theory has been introduced via kernel functors, i.e. O. Goldman's set up, because the notion of an idempotent filter of left ideals could only work right for (pre) sheaves if a suitable space of "underlying points" could be defined, but this seems to fail. It is remarkable though that the filter of a kernel functor is of some use, especially in the Pointwise theory. For other background the reader is referred to basic papers by P. Gabriel and A. Grothendieck.

Results in Chapter I, which happen to be new, are due to F. Van Oystaeyen; other chapters result from both authors joint efforts.

CHAPTER I. THE RING THEORETIC SCENERY

I.1. Localization of Rings and Modules

Throughout, R is a ring with unit and R-module means left R-module. The category of R-modules is denoted by R-mod. A left and right ideal of R is said to be an ideal of R. In this preliminary section we gather some properties of localization functors in R-mod. For more detail on this matter the reader may consult [9], [12], [18].

An endofunctor κ in R-mod is said to be an (idempotent) kernel functor if it is a left exact subfunctor of the identity in R-mod, such that $\kappa(M/\kappa(M)) = 0$ for every $M \in$ R-mod. An R-module M is said to be κ-torsion if $\kappa(M) = M$, κ-torsion free if $\kappa(M) = 0$.

The class of κ-torsion modules is closed under taking submodules, homomorphic images, direct sums and extensions. Conversely, any class of R-modules with these properties is the class of κ-torsion modules with respect to a unique kernel functor κ. A similar statement holds for the class of κ-torsion free modules, which is closed under taking submodules, direct products, injective hulls and isomorphic copies.

Let $L(\kappa)$ be the set of left ideals I of R such that R/I is κ-torsion. $L(\kappa)$ is called the (idempotent) filter associated with κ; it has the following properties :

1. If $I,J \in L(\kappa)$ then $I \cap J \in L(\kappa)$.
2. If $I \in L(\kappa)$ and J is a left ideal of R containing I then

$J \in L(\kappa)$.

3. If $I \in L(\kappa)$ and $x \in R$, then there exists a $J \in L(\kappa)$ such that $Jx \subset I$.
4. If $M \in$ R-mod, $x \in M$, then $x \in \kappa(M)$ if and only if there is an $I \in L(\kappa)$ such that $Ix = 0$.
5. Let I,J be left ideals of R with $I \subset J$ and $J \in L(\kappa)$. If J/I is a κ-torsion module, then $I \in L(\kappa)$.
6. If I is a left ideal of R, $r \in R$, let $(I : r)$ be the left ideal $\{x \in R, xr \in I\}$ of R.

   If for some $J \in L(\kappa)$, $(I : r) \in L(\kappa)$ for every $r \in J$, then $I \in L(\kappa)$.

Idempotent filters are in fact characterized by 3. and 6., or also by 1. to 4. Taking $L(\kappa)$ as a fundamental set of neighborhoods of 0 in R, makes R into a linear topological ring and any linear topological ring structure of R is derived in this way from a kernel functor.

The class of kernel functors on R-mod is a set and it will be denoted by R-ker. This set may be partially ordered as follows : if $\kappa,\kappa' \in$ R-ker then $\kappa \leqslant \kappa'$ if and only if $L(\kappa) \subset L(\kappa')$, or equivalently, if and only if $\kappa(M) \subset \kappa'(M)$ for every $M \in$ R-mod. If $\kappa,\kappa' \in$ R-ker then $\kappa \wedge \kappa'$ is defined to be the kernel functor associated to $L(\kappa) \cap L(\kappa')$, while

$$\kappa \vee \kappa' = \wedge\{\pi \in \text{R-ker}, \pi \geqslant \kappa, \pi \geqslant \kappa'\}.$$

In this way R-ker is a complete distributive lattice.

Let $\kappa \in$ R-ker, $E \in$ R-mod. E is said to be <u>κ-injective</u> if any diagram :

$$\begin{array}{ccccccccc} 0 & \to & M' & \to & M & \to & M/M' & \to & 0 \\ & & f\downarrow & \swarrow_{g} & & & & & \\ & & E & & & & & & \end{array}$$

with exact top row and $\kappa(M/M') = M/M'$, may be completed by g to make it commutative. If g is unique as such then E is said to be <u>faithfully κ-injective</u>. Note that an injective module E is κ-injective for every $\kappa \in$ R-ker.

<u>PROPOSITION</u> I.1.1. Let $E \in$ R-mod. The following statements are equivalent :

1. E is faithfully κ-injective
2. E is κ-injective and κ-torsion free.

<u>PROPOSITION</u> I.1.2. For $E \in$ R-mod, $\kappa \in$ R-ker, the following statements are equivalent :

1. E is κ-injective
2. For every R-homomorphism $f : I \to E$, with $I \in L(\kappa)$, there exists an R-homomorphism $g : R \to E$ such that $g|I = f$.

<u>PROPOSITION</u> I.1.3. Let $0 \to M' \to M \to M'' \to 0$ be an exact sequence in R-mod, then :

1. If M is κ-injective and if M" is κ-torsion free then M' is κ-injective
2. If M' is κ-injective and if M" is κ-torsion then the sequence splits : furthermore if M is κ-torsion free then $M = M'$.

Assume that $M \in$ R-mod is κ-torsion free. Then there exists a faithfully κ-injective module $E_\kappa(M)$ containing M, such that $E_\kappa(M)/M$ is κ-torsion. $E_\kappa(M)$ is unique up to isomorphism in R-mod and it is an essential extension of M.

For an arbitrary $M \in R$-mod define $Q_\kappa(M)$ to be the R-module $E_\kappa(M/\kappa(M))$. This module together with the canonical R-morphism $j_{\kappa,M} : M \to Q_\kappa(M)$ is called the <u>module of quotients of M at κ</u>.

<u>PROPOSITION</u> I.1.4. $Q_\kappa(-)$ is a left exact functor in R-mod.

In particular, $Q_\kappa(R)$ is an R-module; moreover :

<u>PROPOSITION</u> I.1.5. $Q_\kappa(R)$ is a ring containing $R/\kappa(R)$ as a subring. The ring structure in $Q_\kappa(R)$ is determined by its R-module structure and it is unique as such. The canonical R-module homomorphism $j_{\kappa,R} : R \to Q_\kappa(R)$, is a ring homomorphism with kernel $\kappa(R)$. If M is a faithfully κ-injective module then its R-module structure may be extended, in a unique way, to give M the structure of a left $Q_\kappa(R)$-module.

The ring $Q_\kappa(R)$, together with the canonical ring morphism $j_\kappa = j_{\kappa,R} : R \to Q_\kappa(R)$ is called the <u>ring of quotients of R at κ</u>. For every $M \in R$-mod we have that $Q_\kappa(M) \in Q_\kappa(R)$-mod.

The following relates to the flatness of $Q_\kappa(R)$ as an R-module.

<u>PROPOSITION</u> I.1.6. The following statements are equivalent for $\kappa \in R$-ker :

1. Every $M \in Q_\kappa(R)$-mod is κ-torsion free.
2. For any $I \in L(\kappa)$ we have $Q_\kappa(R)j_\kappa(I) = Q_\kappa(R)$.
3. Every $M \in Q_\kappa(R)$-mod is faithfully κ-injective as an R-module.
4. For any $M \in R$-mod, $Q_\kappa(M) \cong Q_\kappa(R) \otimes_R M$.

5. The functor Q_κ is exact and commutes with direct sums.

A kernel functor having these equivalent properties is said to be a <u>t-functor</u> (t for tensor). If $\kappa \in R\text{-ker}$ is a t-functor then $j_\kappa : R \to Q_\kappa(R)$ is (left) flat and an epimorphism in the category of rings. Furthermore :

<u>PROPOSITION</u> I.1.7. If $R \to R'$ is a left flat epimorphism in the category of rings then there exists a unique $\kappa \in R\text{-ker}$ such that $R' \cong Q_\kappa(R)$; moreover : κ is a t-functor.

In case κ is a t-functor, describing $Q_\kappa(R)$-ker is fairly easy because of the following :

<u>PROPOSITION</u> I.1.8. If $\kappa \in R\text{-ker}$ is a t-functor then there is a one-to-one correspondence between the sets $Q_\kappa(R)$-ker and $\{\pi \in R\text{-ker}, \pi \geqslant \kappa\} = \text{gen}(\kappa)$. If $\pi_\kappa \in Q_\kappa(R)$-ker corresponds to $\pi \in \text{gen}(\kappa)$, then $L(\pi_\kappa) = \{Q_\kappa(I), I \in L(\pi)\}$.

We dig a little deeper into the correspondence between R-mod and $Q_\kappa(R)$-mod, for general $\kappa \in R\text{-ker}$.

<u>LEMMA</u> I.1.9. Let $M, M' \in Q_\kappa(R)$-mod. If $f : M \to M'$ is an R-homomorphism such that $Q_\kappa(R)f(M)$ is κ-torsion free as an R-module then f is in fact $Q_\kappa(R)$-linear.

<u>PROOF</u>. Pick $x \in M$, $\lambda \in Q_\kappa(R)$, and let $I \in L(\kappa)$ be such that $I\lambda \subset j_\kappa(R)$. By R-linearity of f : $f(I\lambda x) = If(\lambda x) = (I\lambda)f(x)$. Hence, $f(\lambda x) - \lambda f(x) \in \kappa(Q_\kappa(R)f(M)) = 0$, what proves that f is a

$Q_\kappa(R)$-homomorphism. ■

PROPOSITION I.1.10. Suppose that $M \in Q_\kappa(R)$-mod is κ-torsion free as an R-module, then the following conditions are equivalent :

1. M is injective in R-mod
2. M is injective in $Q_\kappa(R)$-mod.

PROOF. 1. ⇒ 2. Let E be the injective hull of M in $Q_\kappa(R)$-mod. The exact sequence $0 \to M \to E \to E/M \to 0$ splits in R-mod, hence there is an R-homomorphism $f : E \to M$ such that $f|M = 1_M$. The foregoing lemma states that f is a $Q_\kappa(R)$-homomorphism and therefore the exact sequence under consideration, splits in $Q_\kappa(R)$-mod, which entails that $E = M$.

2. ⇒ 1. Let E be the injective hull of M in R-mod. Since M is κ-torsion free, E is κ-torsion free. Hence E is faithfully κ-injective, i.e. $Q_\kappa(E) = E$ or E is a $Q_\kappa(R)$-module. Lemma I.1.9. entails that the inclusion $M \to E$ is $Q_\kappa(R)$-linear, so by 2. the sequence $0 \to M \to E \to E/M \to 0$ splits in $Q_\kappa(R)$-mod. This means that M is a direct summand of E in R-mod and thus $E = M$. ■

COROLLARIES. 1. R-homomorphisms between faithfully κ-injective R-modules are $Q_\kappa(R)$-homomorphisms.

2. If $M \in Q_\kappa(R)$-mod is κ-torsion free as an R-module then the injective hulls of M in R-mod and $Q_\kappa(R)$-mod are isomorphic in R-mod.

I.2. Special Kernel Functors

If R is a commutative ring, S a multiplicatively closed subset of R, then $L(S) = \{\text{ideals } I \text{ of } R,\ I \cap S \neq \phi\}$ is an idempotent filter. Let κ_S be the kernel functor in R-mod associated to $L(S)$, then κ_S is a t-functor. This is not necessarily so when R is non-commutative. This section is built around some basic properties of localization of left Noetherian rings.

R is left Noetherian throughout this section. One associates an idempotent filter to a multiplicatively closed subset S of R, which is supposed not to contain 0 to avoid triviality.
$L(S) = \{I \text{ left ideal of } R,\ (I : r) \cap S \neq \phi \text{ for all } r \in R\}$. The element of R-ker corresponding to $L(S)$ is denoted by κ_S. S is said to be a left Ore set in R if, for given $r \in R$, $s \in S$, there exist $r' \in R$, $s' \in S$ such that $s'r = r's$.

PROPOSITION I.2.1. If S is a left Ore set in the left Noetherian ring R then $\{Rs,\ s \in S\}$ is a filterbasis for $L(S)$ and κ_S is a t-functor.

If P is a prime ideal of R then the complement R-P of P in R is not necessarily multiplicatively closed, but it is an m-system, i.e. if $s_1, s_2 \in R\text{-}P$ then there is an $x \in R$ such that $s_1 x s_2 \in R\text{-}P$. On the other hand the set $G(P) = \{g \in R,\ rg \in P \text{ implies } r \in P\}$ is multiplicatively closed and $\kappa_{G(P)}$ is usually denoted by κ_P. A prime ideal P such that G(P) is a left Ore set is called a localizable prime ideal (sometimes, in this situation, classical prime ideal). Let Q_P denote the localization functor at κ_P.

PROPOSITION I.2.2. If P is a localizable prime ideal of the left Noetherian ring R, then :

1. κ_P is a t-functor.
2. $Q_P(R) j_P(P) = Q_P(P)$ is the Jacobson radical of $Q_P(R)$, where j_P is the canonical ring homomorphism $R \to Q_P(R)$.

3. Elements of $G(P)$ are mapped onto invertible elements of $Q_P(R)$ under j_P.
4. $Q_P(R/P)$ is a simple Artinian ring and it equals the classical ring of quotients $Q_{cl}(R/P)$ of R/P.

PROOF. cf. [19].

If $M \in R$-mod, then it is possible to associate to M two uniquely determined kernel functors of R-ker as follows :

1. Let ξ_M be the minimal element of R-ker for which M is a ξ_M-torsion module.
2. Let χ_M be the maximal element of R-ker for which M is χ_M-torsion free.

Among the obvious and well known properties of ξ_M and χ_M we mention the following. If $N \in R$-mod, then

$\chi_M(N) = \cap\{\text{Ker } f, f : N \to E(M)\}$ where $E(M)$ is an injective hull of M.

Let $\kappa \in R$-ker, A a left ideal of R. A is said to be κ-critical if A is maximal amongst left ideals of R not in $L(\kappa)$. If $\kappa = \chi_{R/A}$ for some κ-critical left ideal A of R then κ is said to be a prime kernel functor. The kernel functor κ_P associated to the prime ideal P is a prime kernel functor.

PROPOSITION I.2.3. Let $\kappa \in R$-ker, then equivalently :

1. κ is prime.
2. $\kappa = \chi_{R/A}$ for every κ-critical left ideal A of R.
3. $\kappa = \chi_M$ for some κ-torsion free $M \in R$-mod such that for every $0 \neq M' \subset M$, M/M' is κ-torsion.

For more details on prime kernel functors, cf. [9] , [12].

PROPOSITION I.2.4. Let $\kappa \in$ R-ker, then $\kappa = \wedge\{\chi_{R/A}$, A is κ-critical$\}$.

A $\kappa \in$ R-ker is said to be symmetric if $L(\kappa)$ has a filterbasis consisting of ideals of R. If $\kappa, \kappa' \in$ R-ker are symmetric then $\kappa \wedge \kappa'$ is symmetric and $\kappa \vee \kappa'$ is symmetric if R is left Noetherian (but not necessarily otherwise). To an arbitrary $\kappa \in$ R-ker one may associate the largest symmetric kernel functor smaller than κ, it is given by $L(\overset{\circ}{\kappa}) = \{I \in L(\kappa)$, I contains an ideal J of R such that $J \in L(\kappa)\}$ and it is denoted by $\overset{\circ}{\kappa}$. Put $\overset{\circ}{\kappa}_P = \kappa_{R-P}$.

PROPOSITION I.2.5. κ_{R-P} is associated to the idempotent filter
$L(\kappa_{R-P}) = \{I$ left ideal of R, $I \supset RsR$ for some $s \in R-P\}$.

Symmetric kernel functors correspond one-to-one to sets of prime ideals of R; to a symmetric $\kappa \in$ R-ker there corresponds the set of ideals maximal with the property of not being in $L(\kappa)$, and to a set of prime ideals $\{P_i, i \in A\}$ there corresponds $L(\kappa) = \{$left ideals J of R containing an ideal I of R such that $I \not\subset P_i$ for all $i \in A\}$. In this way, κ_{R-P} corresponds to $\{P\}$.

If I is an ideal of R then $\chi_{R/I}$ is easily seen to be a symmetric kernel functor and its corresponding idempotent filter has a filterbasis consisting of powers of I.

All symmetric prime kernel functors are of the form κ_{R-P} for some prime ideal P of R. It is a challenging ring-theoretic problem in torsion theory, to find classes of rings for which κ_P, or κ_{R-P}, is a t-functor for all prime ideals P of R. Some examples will arise in sections I.4. and I.5. of this paper.

I.3. Construction of Certain (Pre) Sheaves of Rings and Modules

This section is concerned with the construction of sheaves over spectra of a left Noetherian ring. In particular results of [22], [31] on the prime spectrum will be considerably generalized because here the ring is not supposed to be a prime ring. We include some undispensible definitions here but for the categorical background the reader is referred to Chapter II.

Let X be a topological space and let Open (X) be the category derived from X in the usual way, i.e. using inclusions of open sets for the morphisms. Let $\underline{C}$ be any category. A presheaf of $\underline{C}$-objects over X is a contravariant functor P : open (X) → $\underline{C}$. Let P_V^U be the $\underline{C}$-morphism corresponding to the inclusion $V \subset U$. These $\underline{C}$-morphism are called the restriction maps of P. Let $x \in X$. The stalk of P at x is defined to be $P_x = \varinjlim_{x \in U} P(U)$ in $\underline{C}$. Usually $\underline{C}$ will be an abelian category with exact direct limits, so that the constructed objects we talk about, exist, and have nice properties. If $\underline{C}$ is R-mod then P is a presheaf of R-modules, if $\underline{C}$ is Rings then P is a presheaf of rings. A presheaf P is said to be separated if it has the following property : if $U \in$ Open (X) is covered by $\{U_i,\ i \in A,\ U_i \in \text{Open } (X)\}$ and if $g \in P(U)$ is such that $P_{U_i}^U(g) = 0$ for all $i \in A$, then $g = 0$.

A separated presheaf is said to be a sheaf if and only if : for every $U \in$ Open (X), every open covering $\{U_i,\ i \in A\}$ of U and any set $\{g_i \in P(U_i),\ i \in A,\ P_{U_i \cap U_j}^{U_i}(g_i) = P_{U_i \cap U_j}^{U_j}(g_j)\}$ there exists a (unique) $g \in P(U)$ such that $P_{U_i}^U(g) = g_i$ for all $i \in A$.

If P is a presheaf of R-modules or rings, then to P we may associate a sheaf $\underline{\underline{a}}P$ such that the stalks of P and $\underline{\underline{a}}P$ at x coincide, for all $x \in X$. $\underline{\underline{a}}P$ is called the sheaf associated to P.

Let P,P' be presheaves of $\underline{C}$-objects, then a presheaf morphism $P \to P'$ is defined by giving $\underline{C}$-morphisms $P(U) \to P'(U)$ for every $U \in$ Open (X), which are compatible with the restriction maps P_V^U and P'^U_V, for $V \subset U$ in Open (X). There is a canonical morphism of presheaves $P \to \underline{\underline{a}}P$, and this is a monomorphism, i.e. $P(U) \to \underline{\underline{a}}P(U)$ is monomorphism for all $U \in$ Open (X), if and only if P is a separated presheaf. The $\underline{C}$-object $\underline{\underline{a}}P(U)$, $U \in$ Open (X), is referred to as the $\underline{C}$-object of sections over U, while in case $U = X$, $\underline{\underline{a}}P(X)$ is called the $\underline{C}$-object of global sections. The representation problem may be formulated as follows : given an object Γ of $\underline{C}$, find a topological space $X(\Gamma)$ and a sheaf $S(\Gamma)$ of $\underline{C}$-objects over $X(\Gamma)$ such that $S(\Gamma)(X) \cong \Gamma$. A very well known example of this is the structure sheaf over Spec R for a commutative ring R. In the sequel we solve the problem for a prime left Noetherian ring, although the solution is not functorial as it is in the commutative case.

CASE 1 : Spec R

In this case we assume that R is a left Noetherian ring. Let X = spec R be the set of prime ideals of R and put $X_I = \{P \in \text{spec } R,\ P \not\supset I\}$, where I is an ideal of R. The Zariski topology of spec R is defined by taking the sets X_I for the open sets. To X_I we associate the kernel functor $\kappa_{(I)} = \xi_{R/I}$ which is symmetric; this association is well defined because both X_I and $\kappa_{(I)}$ only depend on the radical rad $I = \cap\{P \in \text{spec } R,\ P \supset I\}$. It is easily verified that a presheaf of rings Q may be defined by putting $Q(X_I) = Q_I(R)$, where Q_I is the localization functor corresponding to $\kappa_{(I)}$. The set spec R equipped with the Zariski topology and the sheaf $\underline{\underline{a}}Q$ is denoted by Spec R, in other words, Spec R is a ringed space in the sense of [13].

PROPOSITION I.3.1. Q is a separated presheaf.

PROOF. It is obvious that the Zariski topology is a T_0-topology and that every open set is quasi-compact. Therefore, open coverings may be supposed to be finite. So let $\{X_{B_i}, 1 \leq i \leq n\}$ be a finite open covering of X_B, where B, B_i are ideals of R, and let κ_i correspond to B_i. Put $Q_i = Q_{B_i}$. If $g \in Q_B(R)$ has the property that its image g_i in $Q_i(R)$ is equal to zero for each i, then take $I \in L(\kappa_{(B)})$ such that $Ig \subset R/\kappa_{(B)}(R)$. Hence $Ig \subset \kappa_i(R)/\kappa_{(B)}(R)$, $i = 1,\ldots,n$. Take $r \in \bigcap_i \kappa_i(R)$; for each i there is an ideal $C_i \in L(\kappa_i)$, such that $C_i x = 0$. By definition of κ_i, rad $C_i \supset B_i$. Take $C = \sum_i C_i$, then rad $C \supset \text{rad}(\sum_i B_i)$. Since $X_B = \cup X_{B_i}$, it follows easily that every $P \in X$ containing $\sum_i$ rad B_i is in $X - X_B$. Hence rad $B \subset \text{rad}(\sum_i B_i)$, which yields that $C \in L(\kappa_{(B)})$. From $Cr = 0$, $r \in \kappa_{(B)}(R)$ follows. Finally, $Ig = 0$ contradicts the fact that $Q_B(R)$ is $\kappa_{(B)}$-torsion free, unless $g = 0$. ■

PROPOSITION I.3.2. If R is a left Noetherian prime ring then $Q \cong \underline{\underline{a}}Q$ as presheaves and $R = \underline{\underline{a}}Q(X)$.

PROOF. cf. [30]. ■

An $X_{(I)} \in$ Open (X) is said to be a geometric t-set if $\kappa_{(I)}$ is a t-functor and if for any ideal B of R we have that $Q_I(R)j_I(B)$ is an ideal of $Q_I(R)$, where J_I is the canonical ring morphism $R \to Q_I(R)$. Recall from [30] :

PROPOSITION I.3.3. Let R be a left Noetherian prime ring and let X_I be a geometric t-set in Spec R then X_I, with the sheaf induced by Spec R is sheaf-isomorphic to Spec $Q_I(R)$. Moreover (geometric) t-sets of Spec $Q_I(R)$

correspond one-to-one to (geometric) t-sets of Spec R which are contained in X_I.

REMARK. The same statements hold if the assumption that R be a prime ring is omitted : the proof goes in exactly the same way.

If $M \in$ R-mod, then let Q(M) be the presheaf of R-modules defined by putting $Q(M)(X_I) = Q_I(M)$, for every ideal I of R. If we assume that M is "absolutely" torsion free, i.e. M is κ-torsion free for all symmetric kernel functors $\kappa \in$ R-ker, then Q(M) is a sheaf and $Q(M)(X) = M$. In general, if $\underline{\underline{a}}Q(M)$ denotes the sheaf associated to to Q(M), the following results generalize to the case of arbitrary left Noetherian rings some theorems which appeared in [23], [30]:

THEOREM I.3.4. Let R be a left Noetherian ring and let M be a finitely generated R-module. The stalk of $\underline{\underline{a}}Q(M)$ at $P \in$ spec R equals $Q_{R-P}(M)$.

PROOF. The stalk of $\underline{\underline{a}}Q(M)$ at P is by definition :

$$Sp(M) = \varinjlim_{P \in X_I} Q_I(M).$$

If $P \in X_I$ then $P \not\supset I$ and thus $\kappa_I \leqslant \kappa_{R-P}$. From this it follows that there exist canonical R-module homomorphisms $Q_I(M) \to Q_{R-P}(M)$. The universal property of $\varinjlim$ implies that we have an R-module homomorphism $s_P : S_P(M) \to Q_{R-P}(M)$. Since R is left Noetherian, every $\kappa \in$ R-ker is of finite type, in particular κ_{R-P} is of finite type. Therefore κ_{R-P} commutes with direct limits, cf. [12], i.e.,

$$\kappa_{R-P}(S_P(M)) = \varinjlim_{P \in X_I} \kappa_{R-P}(Q_I(M)).$$

The following sequence is exact :

$$0 \to \varinjlim_{P \in X_I} \kappa_{R-P}(Q_I(M)) \to S_P(M) \overset{s_P}{\to} Q_{R-P}(M).$$

Let $x \in \kappa_{R-P}(Q_I(M))$, i.e. $Jx = 0$ for some $J \in L(\kappa_{R-P})$. Consider $X_I \cap X_J = X_{IJ}$ and the canonical R-module homomorphism $s_{IJ} : Q_I(M) \to Q_{IJ}(M)$. By R-linearity of s_{IJ} we have $IJs_{IJ}(x) = 0$ and it follows that $s_{IJ}(x) = 0$ because $Q_{IJ}(M)$ is $\kappa_{(IJ)}$-torsion free. Thus, any $x \in \kappa_{R-P}(Q_I(M))$, for any ideal I of R, maps to zero in the direct limit, what proves that s_P is a monomorphism. Now let $x \in Q_{R-P}(M)$; then $Ix \subset M/\kappa_{R-P}(M)$ for some $I \in L(\kappa_{R-P})$ and since κ_{R-P} is symmetric we may suppose that $IRx \subset M/\kappa_{R-P}(M)$. Moreover, M being finitely generated as a left R-module and R being a left Noetherian ring, we may conclude that $\kappa_{R-P}(M)$ is finitely generated and that it can be annihilated by some $I' \in L(\kappa_{R-P})$. Put $J = I \cap I'$. Clearly $J \in L(\kappa_{R-P})$, thus $\kappa_{(J)} \leqslant \kappa_{R-P}$. However, because $J\kappa_{R-P}(M) = 0$, we also have $\kappa_{(J)}(M) = \kappa_{R-P}(M)$. Finally, since $Jx \subset M/\kappa_{R-P}(M)$, we obtain
$x \in Q_J(M/\kappa_{R-P}(M)) = Q_J(M/\kappa_{(J)}(M)) = Q_J(M)$. (Note : we have used the monomorphism $Q_J(M) \to Q_{R-P}(M)$ as an identification of $Q_J(M)$ with a submodule of $Q_{R-P}(M)$.) It follows that s_P is onto and hence an isomorphism. ∎

<u>COROLLARY</u>. If R is a left Noetherian ring, then the stalk of $\underline{\underline{a}}Q$ at $P \in$ spec R is exactly $Q_{R-P}(R)$.

We study a special case where we can get rid of the finiteness condition on M. Spec R is said to have a <u>t-basis</u> if there exists a basis $\{X_B,\ B \in \mathcal{B}\}$ for the Zariski topology on spec R such that the kernel functors $\kappa_{(B)}$, $B \in \mathcal{B}$, are t-functors.

PROPOSITION I.3.5. Let R be a left Noetherian ring such that Spec R has a t-basis, and let $M \in$ R-mod. The stalk of $\underline{\underline{a}}Q(M)$ at $P \in$ spec R equals $Q_{R-P}(M)$.

PROOF. Since R is left Noetherian, the foregoing theorem applied to R yields that $Q_{R-P}(R)$ is the stalk of $\underline{\underline{a}}Q$ at P. If X_I is such that $\kappa_{(I)}$ is a t-functor, then $Q_I(M) \cong Q_I(R) \underset{R}{\otimes} M$. Since taking direct limits commutes with tensor products, we get $S_P(M) \cong Q_{R-P}(R) \underset{R}{\otimes} M$. Hence if we are able to prove that κ_{R-P} is a t-functor then $S_P(M) \cong Q_{R-P}(M)$ follows. Now,

$$\kappa_{R-P} = \vee \{\kappa_{(I)}, I \not\subset P\} = \vee \{\kappa_{(I)}, I \not\subset P \text{ and } \kappa_{(I)} \text{ is a t-functor}\},$$

because the set of $\kappa_{(I)}$ being t-functors determines a cofinal set of neighborhoods of P. It is well known that the supremum of a set of t-functors is a t-functor. ∎

COROLLARY. In the above situation, application of Proposition I.3.3. yields the following : if X_I is a geometric t-set in Spec R, then Spec $Q_I(R)$ has a t-basis.

CASE 2 : R-Sp

Let R-sp be the set of all prime kernel functors in R-ker. If $\kappa \in$ R-ker, then gen$(\kappa) = \{\kappa' \in$ R-ker, $\kappa' \geqslant \kappa\}$ and pgen(κ) = gen$(\kappa) \cap$ R-sp. The topology on R-sp generated by $\{$pgen$(\kappa), \kappa \in C\}$ where $C = \{\xi_{R/I}$, I left ideal of R$\}$ is called the basic order topology. Let X be the topological space thus defined. If $U \in$ Open (X) then we assign to it the kernel functor $\wedge U = \wedge\{\kappa, \kappa \in U\}$. Let $M \in$ R-mod and put $P(M)(U) = Q_{\wedge U}(M)$. P(M) is a presheaf of R-modules which is obviously separated : the sheaf associated to it will be denoted by $\underline{\underline{a}}P(M)$. If $\pi \in X$, then the stalk $S_\pi(M)$ of $\underline{\underline{a}}P(M)$ at π is defined to be $\varinjlim_{U \ni \pi} Q_{\wedge U}(M)$ and we have a canonical R-morphism $s_\pi : S_\pi(M) \to Q_\pi(M)$. R-Sp will be the ringed space on R-sp with its basic order topology.

<u>PROPOSITION</u> I.3.6. For every $\pi \in$ R-sp, S_π is a monomorphism.

<u>PROOF</u>. Let $U = \text{pgen}(\kappa)$ be a neighborhood of π, then $\kappa \leqslant \wedge U \leqslant \pi$ and thus $Q_\pi(M) = Q_\pi Q_{\wedge U}(M)$. This yields an exact sequence :

$$0 \to \pi(Q_{\wedge U}(M)) \to Q_{\wedge U}(M) \to Q_\pi(M).$$

Taking the direct limit over all basic open neighborhoods U of π and using the exactness of $\varinjlim$ we obtain :

$$0 \to \varinjlim \pi(Q_{\wedge U}(M)) \to S_\pi(M) \underset{s_\pi}{\to} Q_\pi(M).$$

If $x \in \pi(Q_{\wedge U}(M))$, then $W = U \cap \text{pgen}(\xi(Rx))$ is a basic open neighborhood of π contained in U such that x maps to zero in $Q_{\wedge W}(M)$. Hence the image of x in $\varinjlim \pi(Q_{\wedge U}(M))$ is 0, therefore s_π is a monomorphism. ∎

<u>PROPOSITION</u> I.3.7. Let R be a left Noetherian ring and let $M \in$ R-mod. Suppose that either one of the following properties holds :

(1) There is a cofinal set C_t in C such that every $\kappa \in C_t$ is a t-functor.

(2) M is a finitely generated R-module.

Then $S_\pi(M) \cong Q_\pi(M)$.

<u>PROOF</u>. The proof of (1) is similar to that of the same statement in case 1 for Spec R.

(2) The foregoing proposition yields that s_π is monomorphic. To show that s_π is epimorphic it will be sufficient to prove that $Q_\pi(M)$ is the union of all images of $\theta_{U,\pi}$,

$$\theta_{U,\pi} : Q_{\wedge U}(M) \to Q_\pi(M).$$

First, $\pi(M)$ is finitely generated because R is left Noetherian and M finitely generated, let $\pi(M) = \sum_{i=1}^{n} Rx_i$. Let $I_i \in L(\pi)$ be such that $I_i x_i = 0$, and then consider ξ_N where $N = \sum_{i=1}^{n} R/I_i$. Obviously $\xi_N(M) = \pi(M)$. Put $I' = \bigcap_{i=1}^{n} I_i$; then $\xi_{R/I'} \leqslant \pi$ and $\xi_{R/I'} \geqslant \xi_N$. If $x \in Q_\pi(M)$ then $I_0 x \subset M/\pi(M)$ for some $I_0 \in L(\pi)$. Put $I = I' \cap I_0$, then $\pi \geqslant \xi_{R/I} \geqslant \xi_{R/I'} \geqslant \xi_N$ while $\xi_{R/I}(M) = \pi(M)$. Therefore $Ix \subset M/\pi(M) = M/\xi_{R/I}(M)$ implies $x \in Q_{\xi_{R/I}}(M/\pi(M)) = Q_{\xi_{R/I}}(M)$. ■

PROPOSITION I.3.8. Let R be a left Noetherian ring and let $\kappa \in$ R-sp be a t-functor, then $Q_\kappa(R)$-Sp $\cong$ pgen(κ), where pgen(κ) is endowed with the induced sheaf, i.e. the restriction of R-Sp to the open set pgen(κ).

The two following sections are mainly concerned with the description of certain classes of left Noetherian rings R for which Spec R $\cong$ R-Sp or for which Spec R has a t-basis.

I.4. Fully Left Bounded Left Noetherian Classical Rings

In this section, R is still assumed to be a left Noetherian ring. A prime ring R is said to be left bounded if every essential left ideal of R contains a non-zero ideal of R. R is said to be fully left bounded if R/P is left bounded for every P $\in$ spec R.

PROPOSITION I.4.1. If R is fully left bounded then every idempotent kernel functor $\kappa \in$ R-ker is symmetric.

PROOF. It is well known, cf. [26], that for a fully left bounded ring, the Gabriel-correspondence between prime ideals of R and indecomposable injective R-modules is one-to-one. For any $\kappa \in$ R-ker, let κ^{o} be the largest symmetric kernel functor smaller than κ; i.e. $L(\kappa^{o})$ = {left ideals of R containing an ideal which is in $L(\kappa)$}. If A is a κ^{o}-critical left ideal then $\kappa_{R/A}$ is a prime kernel functor hence it equals κ_P for some $P \in$ spec R. Since $\kappa^{o} \leq \kappa_{R/A} = \kappa_P$, R/P is κ^{o}-torsion free. Assume that $\kappa(R/P) \neq 0$, then $\kappa(R/P)$ is an ideal of the left Noetherian prime ring R/P. Hence $\kappa(R/P)$ is essential as a left ideal and thus it contains a regular element x of R/P. Let $I \in L(\kappa)$ be such that $Ix = 0$, then $I \subset P$ and $P \in L(\kappa^{o})$ follows. The latter however contradicts $\kappa^{o}(R/P) = 0$, there $\kappa(R/P) = 0$ and $\kappa \leq \kappa_P = \chi_{R/P} = \chi_{R/A}$ follows. Finally, because $\kappa \leq \inf\{\chi_{R/A}$, A being κ^{o}-critical$\} = \kappa^{o}$, we get $\kappa = \kappa^{o}$. ■

Some elementary facts about fully left bounded rings may be summarized as follows :

PROPOSITION I.4.2. The following statements are equivalent :

1. R is fully left bounded
2. The Gabriel-correspondence is one-to-one
3. Every $\pi \in$ R-sp is symmetric
4. Every $\kappa \in$ R-ker is symmetric.

Note that κ_P and κ_{R-P} coincide for all $P \in$ spec R.

PROPOSITION I.4.3. Let R be fully left bounded. The following statements are equivalent :

1. P is a classical prime ideal of R
2. $P \subset A$ for every κ_P-critical left ideal A of R
3. κ_P is a t-functor and $Q_P(P)$ is the Jacobson radical of $Q_P(R)$.

<u>PROOF</u>. Let $C(\kappa_P)$ be the set of κ_P-critical left ideals.
1. ⇒ 2. If P is classical then any $A \in C(\kappa_P)$ is maximal among left ideals not intersecting G(P). If $P \not\subset A$ for some $A \in C(\kappa_P)$ then there is a $g \in G(P)$ such that $g = a + p$ with $a \in A$, $p \in P$. Suppose $ta \in P$ for some $t \in R$, then $tg \in P$ and $t \in P$ follows; thus $a \in G(P)$. This contradicts $A \cap G(P) = \phi$ and thus $P \subset A$ for all $A \in C(\kappa_P)$.
2. ⇒ 1. Pick $g \in G(P)$. If $Rg \notin L(\kappa_P)$ then $Rg \subset A$ for some $A \in C(\kappa_P)$. Since g maps onto a regular element of R/P, and since the latter ring is left bounded, it follows that Rg mod P contains a non-zero ideal of R/P. Therefore $P + Rg \in L(\kappa_{R-P}) = L(\kappa_P)$. However $A \in C(\kappa_P)$ and $A \supset P + Rg$ yields a contradiction, therefore $Rg \in L(\kappa_P)$ for every $g \in G(P)$. By definition of $L(\kappa_P)$, we have $[Rg : r] \cap G(P) \neq \phi$ for every $r \in R$, but this expresses exactly that R satisfies the left Ore condition with respect to G(P).
1. ⇒ 3. cf. [28].
3. ⇒ 2. κ_P being a t-functor, the maximal left ideals of $Q_P(R)$ are exactly the $Q_P(A)$, $A \in C(\kappa_P)$, as is easily seen. Hence $Q_P(P) = \cap \{Q_P(A), A \in C(\kappa_P)\}$, entailing $P = \cap\{A, A \in C(\kappa_P)\}$.

A ring R is said to be <u>left balanced</u> if for critical left ideals I', I of R such that $I' \subset I$, we have that $\chi_{R/I'} \geq \chi_{R/I}$. Details about these rings may be found in [9] or the references there.

<u>PROPOSITION</u> I.4.4. If R is a left Noetherian fully left bounded

classical ring then R is left balanced.

PROOF. If $I' \subset I$ are critical left ideals of R then $\chi_{R/I}$ and $\chi_{R/I'}$ are prime kernel functors and we may write $\chi_{R/I} = \kappa_P$, $\chi_{R/I'} = \kappa_{P'}$. Then I is κ_P-critical and I' is $\kappa_{P'}$-critical. Since P' is a classical prime ideal, Proposition I.4.3. applies and it entails that $P' \subset I' \subset I$. Obviously P is the largest ideal contained in I, (indeed if $J \subset I$ and $J \not\subset P$ then $J \in L(\kappa_{R-P}) = L(\kappa_P)$ would entail $I \in L(\kappa_P)$, contradicting $I \in C(\kappa_P)$.), hence $P' \subset P$ and $\kappa_P \leq \kappa_{P'}$ i.e. $\kappa_{R/I} \leq \kappa_{R/I'}$. ∎

If $C = \{\xi_{(R/I)}$, I left ideal of R)}, put $C^0 = \{\xi_{(R/I)}$, I ideal of R}. The sets pgen(κ), $\kappa \in C$ form a basis for the basic order topology of R-Sp. Recall also that for a left balanced left (semi) Noetherian ring R and $\kappa \in C$ we have that $\kappa = \wedge$ pgen(κ).

THEOREM I.4.5. <u>If</u> R <u>is a left Noetherian fully left bounded classical ring</u>, <u>then</u> R-Sp <u>is homeomorphic to</u> Spec R. <u>Moreover they are isomorphic as ringed spaces</u>.

PROOF. To a left ideal I of R we associate an ideal I^0 of R, which is the biggest ideal of R contained in I. By proposition I.4.1. , $\xi_{(R/I)}$ is symmetric, hence $I^0 \in L(\xi_{(R/I)})$. Therefore R/I^0 is $\xi_{(R/I)}$-torsion and thus $\xi_{(R/I^0)} \leq \xi_{(R/I)}$. Conversely, because of the existence of the R-epimorphism $R/I^0 \to R/I$, we have $\xi_{(R/I)} \leq \xi_{(R/I^0)}$, whence equality follows. Furthermore, $(I^0)^n \in L(\xi_{(R/I^0)})$ for all $n \geq 0$. $\kappa_{(I^0)} \in$ R-ker is given by $L(\kappa_{(I^0)}) = \{J$ left ideal of R, $J \supset (I^0)^n$ for some $n \geq 0\}$. Obviously, $\kappa_{(I^0)} \leq \xi_{(R/I^0)}$. So, since all $\kappa \in C$ are symmetric, it follows that, in defining the basic order topology on R-sp, the set C may be replaced by C^0 without changing the topology. So, let $\xi_{(R/I)} \in C^0$ and let $\pi \in \mathrm{pgen}(\xi_{(R/I)})$. Since π is a prime kernel functor we have $\pi = \chi_{R/P}$ for some $P \in \mathrm{spec}\ R$. By the foregoing proposition, R is left balanced and then $\chi_{R/P} \geqslant \xi_{(R/I)}$ is equivalent to $\xi_{(R/P)} \nleqslant \xi_{(R/I)}$, cf. [24]. Now, if $P \supset I$, then $R/I \to R/P$ is an R-epimorphism, yielding $\xi_{(R/P)} \leqslant \xi_{(R/I)}$. Thus from $\chi_{R/P} \geqslant \xi_{(R/I)}$ we deduce that $P \not\supset I$. The mapping $\chi_{R/P} \to P$ is a bijective map R-sp $\to$ spec R which maps $\mathrm{pgen}(\xi_{(R/I)})$ to X_I In spec R. This yields the desired homeomorphism. Moreover, since corresponding open sets have the same kernel functor associated to them in either Spec or R-Sp, the isomorphism Spec R $\cong$ R-Sp is evident. ■

REMARKS. Rings with non-trivial polynomial identity are fully bounded. Azumaya algebras, cf. [31], or group rings of finite nilpotent groups over a commutative Noetherian ring are Noetherian, fully bounded and classical. Another example of such a ring is a ring R which is a finitely generated module over its Noetherian center C and such that the prime ideals of R have the unique lying-over property, i.e. if $P \cap C = Q \cap C$ for $P,Q \in \mathrm{spec}\ R$ then $P = Q$ follows.

I.5. Zariski Central Rings

Before studying a class of rings R for which Spec R has a t-basis, we introduce compatibility conditions for kernel functors. The notion of compatibility is interesting also because it inspires some definitions in Section II.4. Moreover Zariski central rings have very specific behaviour with respect to compatibility.

Let R be an arbitrary ring with unit and let $\kappa,\kappa' \in$ R-ker. We say that κ' is Q_κ-compatible if the following holds in R-mod : $\kappa' Q_\kappa = Q_\kappa \kappa'$ (natural equivalence).

PROPOSITION I.5.1. If $\kappa', \kappa \in$ R-ker are such that $\kappa' \geqslant \kappa$ then κ' is Q_κ-compatible.

PROOF. Take $M \in$ R-mod. Since $\kappa' \geqslant \kappa$ we have $\kappa'(M) \supset \kappa(M)$ and $L(\kappa') \supset L(\kappa)$. Take $x \in Q_\kappa(\kappa'(M))$ and let $I \in L(\kappa)$ be such that $Ix \subset \kappa'(M)/\kappa(M)$. It is clear that $(\text{Ann } x : r) \in L(\kappa')$ for every $r \in I$, with $I \in L(\kappa) \subset L(\kappa')$. Since $L(\kappa')$ is an idempotent filter it follows that $\text{Ann } x \in L(\kappa')$ and thus $x \in \kappa'(Q_\kappa(M))$. On the other hand $M/\kappa'(M)$ is κ-torsion free and we obtain an essential monomorphism $M/\kappa'(M) \to Q_\kappa(M/\kappa'(M))$. Since $\kappa'(Q_\kappa(M/\kappa'(M))$ intersects $M/\kappa'(M)$ in zero, it has to be the zero module. Left exactness of Q_κ yields that $Q_\kappa(M)/Q_\kappa(\kappa'(M))$ is κ'-torsion free, thus $\kappa' Q_\kappa(M) \subset Q_\kappa(\kappa'(M))$. ■

Let $\rho : R \to R'$ be a ring homomorphism. To ρ there corresponds a functor $\rho_* : R'\text{-mod} \to R\text{-mod}$, the "restriction of scalars". Since ρ_* is exact and commutes with direct sums we may conclude that ρ_* gives rise to a mapping $\overline{\rho} : R\text{-ker} \to R'\text{-ker}$, which is a morphism of complete distributive lattices. If $\kappa \in$ R-ker, then $\overline{\rho}(\kappa)$ may be described by its class of torsion modules, i.e. $M \in R'$-mod is $\overline{\rho}(\kappa)$-torsion if and only if $\rho_*(M)$ is κ-torsion. In this terminology, the definition of Q_κ-compatibility for κ' is : $\rho_* Q_\kappa \kappa' = \kappa' Q_\kappa$ where $\rho = j_\kappa : R \to Q_\kappa(R)$, the canonical ring homomorphism. If $\kappa' \in$ R-ker, then $\overline{j}_\kappa(\kappa')$ will be denoted by κ'_κ.

THEOREM I.5.2. Let $\kappa', \kappa \in$ R-ker and suppose that κ' is Q_κ-compatible, then $j_\kappa^* Q_{\kappa'_\kappa} Q_\kappa = Q_{\kappa'} Q_\kappa$ (natural equivalence of functors).

<u>PROOF</u>. We have to show that $Q_{\kappa'}Q_{\kappa}(M)$ is isomorphic to $Q_{\kappa'_{\kappa}}Q_{\kappa}(M)$ considered as an R-module, for every $M \in$ R-mod. The definition of $\kappa'_{\kappa} \in Q_{\kappa}(R)$-ker guarantees that

$$\kappa'_{\kappa}\, Q_{\kappa}(M) = \kappa' Q_{\kappa}(M) \cong Q_{\kappa}(\kappa'(M)).$$

Put

$$N = Q_{\kappa}(M)/\kappa'(Q_{\kappa}(M)) = Q_{\kappa}(M)/Q_{\kappa}(\kappa'(M)).$$

We have a monomorphism in $Q_{\kappa}(R)$-mod : $0 \to N \to Q_{\kappa}(M/\kappa'(M))$. Moreover N is κ'-torsion free as an R-module because :

$$\kappa'(Q_{\kappa}(M/\kappa'(M)) = Q_{\kappa}(\kappa'(M/\kappa'(M))) = 0.$$

Because of the monomorphism $N \to Q_{\kappa}(M/\kappa'(M))$, N is also κ-torsion free, hence the injective hulls of N in R-mod and $Q_{\kappa}(R)$-mod are isomorphic in R-mod, cf. Corollary 2, Proposition I.1.10. From the construction of $Q_{\kappa'_{\kappa}}(N)$ it follows that we have inclusions $N \subset Q_{\kappa'_{\kappa}}(N) \subset E(N)$, where E(N) is an injective hull of N. By construction again, $E(N)/Q_{\kappa'_{\kappa}}(N)$ is κ'_{κ}-torsion free hence κ'-torsion free as an R-module. Since E(N) is obviously κ'-injective, it follows from Proposition I.1.3., 1., that $Q_{\kappa'_{\kappa}}(N)$ is κ'-injective. But, as $Q_{\kappa'_{\kappa}}(N)/N$ is κ'_{κ}-torsion hence κ'-torsion as an R-module, all this amounts to $Q_{\kappa'}(N) \cong Q_{\kappa'}(N)$ in R-mod. Now, it is clear that $Q_{\kappa'}(N) = Q_{\kappa'_{\kappa}}Q_{\kappa}(M)$ on one hand, while on the other hand : $Q_{\kappa'}(N) = Q_{\kappa'}Q_{\kappa}(M)$. ■

<u>COROLLARY</u>. If $\kappa \in$ R-ker is a t-functor and if κ' is Q_{κ}-compatible then : $(\kappa' \vee \kappa)_{\kappa} = \kappa'_{\kappa}$ and

$$Q_{\kappa' \vee \kappa}(M) \doteq Q_{\kappa'_{\kappa}}Q_{\kappa}(M) = Q_{\kappa'}Q_{\kappa}(M),$$

and it is easily verified that $L(\kappa'_{\kappa}) = \{Q_{\kappa}(I),\ I \in L(\kappa')\}$.

PROPOSITION I.5.3. Let R be a left Noetherian ring, let $\kappa, \kappa' \in$ R-ker and suppose that κ is symmetric. If κ' is Q_κ-compatible then for any $I \in L(\kappa')$, $J \in L(\kappa)$ there exist $I' \in L(\kappa')$, $J' \in L(\kappa)$ such that $J'I' \subset IJ$.

PROOF. Consider the following exact sequence in R-mod :

$$0 \to J/IJ \to R/IJ \to R/J \to 0.$$

The left exactness of Q_κ entails that the following sequence is exact too :

$$0 \to Q_\kappa(J/IJ) \to Q_\kappa(R/IJ) \to Q_\kappa(R/J).$$

Since $J \in L(\kappa)$, $Q_\kappa(R/J) = 0$ and $Q_\kappa(J/IJ) \cong Q_\kappa(R/IJ)$. Since J/IJ is obviously κ'-torsion, the Q_κ-compatibility of κ' yields that $Q_\kappa(J/IJ)$, hence $Q_\kappa(R/IJ)$ is κ'-torsion. So R/IJ mod $\kappa(R/IJ)$ is κ'-torsion. If $e \in R/IJ$ is the image of $1 \in R$ then $I'e \subset \kappa(R/IJ)$ for some $I' \in L(\kappa')$. Because $I'e$ is finitely generated as an R-module and because κ is symmetric, there exists a $J' \in L(\kappa)$ such that $J'I'e = 0$, hence $J'I' \subset IJ$. ■

PROPOSITION I.5.4. Let R be a left Noetherian ring, let $\kappa, \kappa' \in$ R-ker be symmetric, then the following statements are equivalent :

1. κ' is Q_κ-compatible and κ is $Q_{\kappa'}$-compatible
2. For $I \in L(\kappa')$, $J \in L(\kappa)$ there exist $I', I'' \in L(\kappa')$ and $J', J'' \in L(\kappa)$ such that $J'I' \subset IJ$, $I''J'' \subset JI$.

PROOF. 1. ⇒ 2. follows from the foregoing proposition.
2. ⇒ 1. Let $\pi : M \to M/\kappa(M)$ be the canonical epimorphism, $M \in$ R-mod.

Then obviously $\pi(\kappa'(M)) \subset \kappa'(\pi(M))$ and it is also clear that $\pi^{-1}(\kappa'(\pi(M))) \subset (\kappa' \vee \kappa)(M)$. Moreover, by 2. it follows that $L(\kappa \vee \kappa')$ has a filterbasis $\{JI, J \in L(\kappa), I \in L(\kappa')\}$, thus if $x \in (\kappa \vee \kappa')(M)$ then $JIx = 0$ for some $J \in L(\kappa)$, $I \in L(\kappa')$. Hence : $Ix \in \kappa(M)$, or $\pi(x) \in \kappa'(\pi(M))$. This proves that $\pi^{-1}(\kappa'(\pi(M))) = (\kappa' \vee \kappa)(M)$. Furthermore, if $x \in (\kappa' \vee \kappa)(M)$ then $JIx = 0$ yields $I''J''x = 0$ or $J''x \subset \kappa'(M)$, entailing that $(\kappa' \vee \kappa)(M)/\kappa'(M)$ is κ-torsion. Therefore, $\kappa'(\pi(M))/\pi(\kappa'(M))$ is κ-torsion, i.e. $Q_\kappa(\pi(\kappa'(M))) = Q_\kappa(\kappa'(\pi(M))$. Take $x \in Q_\kappa(\kappa'(M))$, then $IJx = 0$ for some $J \in L(\kappa)$ such that $Jx \subset \pi(\kappa'(M))$ and $I \in L(\kappa')$. By 2. it follows again that $J'I'x = 0$ but since $I'x \subset Q_\kappa(M)$ it is κ-torsion free, hence $J'I'x = 0$ yields $I'x = 0$ and $x \in \kappa'Q_\kappa(M)$. Conversely, if $x \in \kappa'Q_\kappa(M)$ then for some $J \in L(\kappa)$, $Jx \subset \kappa'Q_\kappa(M) \cap \pi(M) = \kappa'(\pi(M))$, hence :

$$x \in Q_\kappa(\kappa'(\pi(M))) = Q_\kappa(\pi(\kappa'(M)) = Q_\kappa(\kappa'(M)).$$

Since $Q_\kappa\kappa' = \kappa'Q_\kappa$ in R-mod, κ' is Q_κ-compatible. A symmetry argument finishes the proof. ■

PROPOSITION I.5.5. Let R be a left Noetherian ring, let κ', κ be symmetric elements of R-ker and suppose that κ' is Q_κ-compatible while κ is a t-functor, then κ'_κ is symmetric in $Q_\kappa(R)$-ker.

PROOF. Let $\pi : R \to R/\kappa(R)$ be the canonical epimorphism. If $D \in L(\kappa'_\kappa)$ then $D \supset \pi(I)$ for some ideal I of R, $I \in L(\kappa')$; indeed $\pi(R)/D \cap \pi(R)$ is κ'-torsion, hence $\pi^{-1}(D \cap \pi(R))$ contains an $I \in L(\kappa')$ which may be taken to be an ideal since κ' is symmetric. Given $J \in L(\kappa)$, there exist $J' \in L(\kappa)$ and $I' \in L(\kappa')$ such that $J'I' \subset IJ \subset I$. Put $D' = J'I'$, then $J'D' \subset J'I' \subset IJ$. Consider

$\sum_i q_i d_i q_i'$ with $d_i \in \pi(D')$ and $q_i, q_i' \in Q_\kappa(R)$. Choose J such that $Jq_i' \subset \pi(R)$ for all i, and choose $J'' \in L(\kappa)$ such that $J''q_i \subset J'$ for all i. Then :

$$J'' \sum_i q_i d_i q_i' \subset \sum_i J' d_i q_i' \subset \sum_i IJq_i' \subset \pi(I).$$

Since κ is a t-functor we have that $\sum_i q_i d_i q_i' \in Q_\kappa(R)\pi(I)$, hence : $Q_\kappa(R)\pi(D')Q_\kappa(R) = Q_\kappa(R)\pi(I')Q_\kappa(R)$ is contained in $Q_\kappa(R)\pi(I)$ and thus in D. However $Q_\kappa(R)\pi(I')Q_\kappa(R) \supset Q_\kappa(R)\pi(I') = Q_\kappa(I')$ is in $L(\kappa_\kappa')$, cf. Corollary to Theorem I.5.2., therefore D contains an ideal of $Q_\kappa(R)$ which is in $L(\kappa_\kappa')$, i.e. κ_κ' is symmetric. ■

PROPOSITION I.5.6. Let R be a left Noetherian, fully left bounded ring, let $\kappa \in$ R-ker be a t-functor, then $Q_\kappa(R)$ is a left Noetherian fully left bounded ring.

PROOF. Let $\pi \in Q_\kappa(R)$-ker. By Proposition I.1.8., $\pi = \kappa_\kappa'$ for some $\kappa' \in$ R-ker, $\kappa' \geq \kappa$. Since R is fully left bounded we may use Proposition I.4.1. to conclude that κ and κ' are symmetric. Moreover by Proposition I.5.1. it follows from $\kappa' \geq \kappa$ that κ' is Q_κ-compatible and then the foregoing proposition yields that κ_κ' is symmetric. So, every $\pi \in Q_\kappa(R)$-ker is symmetric, thus $Q_\kappa(R)$ is fully left bounded. ■

Let R be a left Noetherian ring with unit and let C be the center of R. R is said to be a Zariski central ring if the Zariski topology of spec R is generated by the open sets X_I, X = spec R, for which I is an ideal of R such that $I = R(I \cap C)$. Denote this generating set of open sets by B.

LEMMA I.5.7. The following statements are equivalent :

(1) R is Zariski central.

(2) $I \subset \text{rad}(I \cap C)$ for all ideals I of R

(3) $P = \text{rad}\, R(P \cap C)$ for all $P \in \text{spec}\, R$.

PROOF. First note that $\mathcal{B}$ is a basis for a topology. Indeed, if X_I, $X_J \in \mathcal{B}$ then $X_I \cap X_J = X_{IJ}$ and it is clear that

$$R(I \cap C)(J \cap C) = I(J \cap C) = IR(J \cap C) = IJ.$$

(1) $\Leftrightarrow$ (2). Let I be an ideal of R, then (1) yields that $X_I = \cup X_i$ where X_i is the open set associated to $R(I_i \cap C)$ for some ideal I_i of R. Now, $X_I \supset X_i$ yields $\text{rad}\, I \supset R(I_i \cap C)$ and $R(\text{rad}\, I \cap C) \supset R(I_i \cap C)$ for all i. Since $X_I \subset X_{R(I \cap C)}$ follows, we also have $\text{rad}\, I = \text{rad}(I \cap C)R$. Conversely, (2) yields $X_I = X_{R(I \cap C)}$ for every ideal I of R.

(2) $\Rightarrow$ (3). Obvious

(3) $\Rightarrow$ (2). There is a finite number of minimal prime ideals P_i containing I, $\text{rad}\, I = \bigcap_i P_i$. Since $\text{rad}\, I \cap C = \bigcap_i (P_i \cap C)$ we have that, if $P \in \text{spec}\, R$, $P \supset \text{rad}\, I \cap C$, then $P \cap C = p$ is a prime ideal of C containing $\bigcap_i p_i$, where $p_i = P_i \cap C$. Hence $p \supset p_i$ and thus $\text{rad}\, R_p \supset \text{rad}\, R_{p_i}$, what yields $P \supset P_i$ for some i. Therefore $P \supset I$ if and only if $P \supset \text{rad}\, I \cap C$, i.e. $\text{rad}\, I = \text{rad}\, R(\text{rad}\, I \cap C)$. Clearly, for some positive integer n, $[R(\text{rad}\, I \cap C)]^n \subset I$; this immediately proves (2). ■

LEMMA I.5.8. If I is an ideal of a Zariski central ring R then R/I is Zariski central.

PROOF. If $\bar{J}$ is an ideal of $\bar{R} = R/I$, let J be $\pi^{-1}(\bar{J})$ where π is the canonical ring epimorphism $R \to R/I$. Since J is an ideal of R, $J^n \subset R(J \cap C)$ for some $n \geq 0$. Therefore

$\overline{J}^n \subset \overline{R}\pi(J \cap C) \subset \overline{R}(\overline{J} \cap \pi C)$ and since πC is in the center of $\overline{R}$, this finishes the proof. ■

<u>PROPOSITION</u> I.5.9. Let R be Zariski central and let κ, κ' be symmetric kernel functors, then κ' is Q_κ-compatible and κ is $Q_{\kappa'}$-compatible.

<u>PROOF</u>. Take ideals $I \in L(\kappa')$, $J \in L(\kappa)$. Since $L(\kappa')$ and $L(\kappa)$ are multiplicatively closed we have also that $R(I \cap C) \in L(\kappa')$, $R(J \cap C) \in L(\kappa)$. Therefore, for left ideals $I' \in L(\kappa')$ and $J' \in L(\kappa)$ we have that there exist $I_1 \in L(\kappa')$ and $I_1 \in L(\kappa)$ such that $J_1 I_1 \subset I'J'$. A similar result holds if I is replaced by J everywhere because of the symmetry in κ and κ'. Therefore, Proposition I.5.4., yields the desired compatibility. ■

If $\kappa \in$ R-ker is symmetric, R a Zariski central ring with center C, then we may define $\kappa^C \in$ C-ker by the filter $L(\kappa^C)$, generated by the set $\{I \cap C, I \in L(\kappa)\}$, which is easily verified to be an idempotent filter. If $j : C \to R$ is the canonical inclusion, then $\overline{j}(\kappa) = \kappa^C$, cf. remarks before Theorem I.5.2..

<u>THEOREM</u> I.5.10. <u>Let</u> R <u>be a Zariski central ring and let</u> $\kappa \in$ R-ker <u>be such that</u> κ^C <u>is a</u> t-<u>functor</u>, <u>then</u> :

1. $Q_\kappa(R)$ <u>is ring-isomorphic to</u> $Q_{\kappa^C}(R)$.
2. $Q_\kappa(M) \cong Q_{\kappa^C}(M)$ <u>in</u> C-<u>mod</u>, <u>for every</u> $M \in$ R-<u>mod</u>.

<u>PROOF</u>. Since κ^C is a t-functor :

$$Q_{\kappa^C}(R) \cong Q_{\kappa^C}(C) \underset{C}{\otimes} R \cong R \underset{C}{\otimes} Q_{\kappa^C}(C)$$

and this is a ring. The C-linear map $j_{\kappa^C} : R \to Q_{\kappa^C}(R)$ is a ring morphism which is easily seen to be left flat, i.e. $Q_{\kappa^C}(R)$ is flat as a right R-module because of the above isomorphisms.

Consider :

$$R \xrightarrow[j_{\kappa^C}]{} Q_{\kappa^C}(R) \overset{f}{\underset{g}{\rightrightarrows}} R_1 \quad ,$$

where R_1 is a ring and f,g are ring morphisms such that $fj_{\kappa^C} = gj_{\kappa^C}$. Pick $x \in Q_{\kappa^C}(R)$, then $Ix \subset R/\kappa^C(R)$ for some $I \in L(\kappa^C)$ i.e. $f(Ix) = g(Ix)$. Now, $C \to Q_{\kappa^C}(C)$ is an epimorphism in the category of rings, where $fj_{\kappa^C}|C = gj_{\kappa^C}|C$ and this yields that $f|Q_{\kappa^C}(C) = g|Q_{\kappa^C}(C)$. Hence, R_1 becomes equipped with an unambiguous $Q_{\kappa^C}(C)$-module structure via either $f|Q_{\kappa^C}(C)$ or $g|Q_{\kappa^C}(C)$. Since κ^C is a t-functor, R_1 is κ^C-torsion free as a C-module, the C-module structure being induced by the $Q_{\kappa^C}(C)$-module structure. However $f(I)f(x) = g(I)g(x)$ yields $I(f(x) - g(x)) = 0$ in the C-module structure of R_1, i.e., $f(x) - g(x) = 0$. This proves that $f = g$ and j_{κ^C} is an epimorphism in the category of rings. It is well known that to such a left flat ring epimorphism there corresponds a t-functor $\kappa' \in$ R-ker such that $Q_{\kappa'}(R) = Q_{\kappa^C}(R)$. Clearly :

$$Q_{\kappa'}(M) \cong Q_{\kappa'}(R) \underset{R}{\otimes} M \cong Q_{\kappa^C}(R) \underset{R}{\otimes} M \cong$$

$$\cong Q_{\kappa^C}(C) \underset{C}{\otimes} R \underset{R}{\otimes} M \cong Q_{\kappa^C}(C) \underset{C}{\otimes} M \cong Q_{\kappa^C}(M),$$

for every $M \in$ R-mod.

If $M \in$ R-mod is κ'-torsion then $Q_{\kappa'}(M) \cong Q_{\kappa^C}(M) = 0$ thus M is κ^C-torsion, hence κ-torsion as an R-module. Conversely, if M is κ-torsion, then it is κ^C-torsion as a C-module hence κ'-torsion; therefore $\kappa' = \kappa$. ∎

<u>COROLLARY</u> 1. In the situation of the foregoing theorem, κ is a t-functor whenever κ^C is.

In particular, taking $\kappa = \kappa_{R-P}$ for some prime ideal P of R, one finds that $\kappa^c = \kappa_{C-p}$, $p = P \cap C \in \text{spec } C$. Since C is commutative, κ^c is a t-functor, so κ is a t-functor. So, as a result, κ_{R-P} is a t-functor for every $P \in \text{spec } R$.

COROLLARY 2. If R is Zariski central and finitely generated as a C-module then every $P \in \text{spec } R$ is classical i.e. R satisfies the left Ore condition with respect to G(P) and P satisfies the Artin-Rees condition. This follows immediately from [25], since we have proved that κ_{R-P} is in fact a central localization. Azumaya algebras are therefore classical rings. Conversely, if R is finitely generated over its Noetherian center, classical, and such that every $P \in \text{spec } R$ is maximal, then R is Zariski central, cf. [28].

PROPOSITION I.5.11. Let R be a Zariski central ring, κ a symmetric kernel functor such that κ^c is a t-functor, then for every ideal I of R, $Q_\kappa(I)$ is an ideal of $Q_\kappa(R)$.

PROOF. $Q_\kappa(R)$ is a central extension of $R/\kappa(R)$, by the foregoing theorem. Moreover, since κ is a t-functor, $Q_\kappa(I) = Q_\kappa(R)j_\kappa(I)$, thus it is an ideal of $Q_\kappa(R)$. ■

COROLLARY. Every t-set in Spec R is a geometric t-set.

PROPOSITION I.5.12. If R is a Zariski central ring then Spec R has a t-basis.

PROOF. Since R is Zariski central the set
$\mathcal{B} = \{X_I,\ I = R(I \cap C),\ I \text{ an ideal of } R\}$ is a basis for the Zariski topology of Spec R. It is clear that Spec C has a t-basis, because C is commutative. To a $\kappa^c \in$ C-ker which is associated to a

t-set in the basis for the Zariski topology on Spec C, there corresponds a $\kappa \in R\text{-ker}$ which is a t-functor, and which is associated to an open set of $\mathcal{B}$. Therefore $\mathcal{B}(t) = \{X_I,\ I = R(I \cap C) \text{ and } \kappa_{(I)} \text{ is a t-functor}\}$ is a basis for the Zariski topology of Spec R. ■

PROPOSITION I.5.13. Let R be a Zariski central ring, $\kappa \in R\text{-ker}$ a t-functor, then $Q_\kappa(R)$ is Zariski central.

PROOF. Since R is Zariski central, so is $R/\kappa(R)$. Put $\kappa' = \bar{j}_\kappa(\kappa)$; then $Q_{\kappa'}(R/\kappa(R)) = Q_\kappa(R)$. So in proving the statement we may assume that R is κ-torsion free. If I is an ideal of $Q_\kappa(R)$, then $I \cap R$ is an ideal of R, so $(I \cap R)^n \subset (I \cap C)R$ for some $n \geq 0$. Hence :

$$Q_\kappa(R)(I \cap R)^n \subset Q_\kappa(R)(I \cap C) \subset Q_\kappa(R)(I \cap Q_{\kappa^C}(C)).$$

Since κ is a t-functor we have that $Q_\kappa(R)(I \cap R) = I$ and since I is an ideal of $Q_\kappa(R)$, $Q_\kappa(R)(I \cap R)^n = I^n$. Therefore, $I^n \subset Q_\kappa(R)(I \cap Q_{\kappa^C}(C))$, where $Q_{\kappa^C}(C)$ is the center of $Q_\kappa(R)$, which entails that $Q_\kappa(R)$ is Zariski central.

COROLLARY. If X_I is a t-set in Spec R, then it is a geometric t-set by the corollary to Proposition I.5.11. Therefore, Proposition I.3.3. yields that $X_I \cong \text{Spec } Q_I(R)$ as ringed spaces. So, for Zariski central rings R, Spec R is a "variety" in the sense that Spec R may be covered by "affine varieties", i.e. geometric t-sets.

PROPOSITION I.5.14. Let R be a left Noetherian ring with center C which is a semilocal ring. If $Q_{C-p}(R)$ is Zariski central for every maximal ideal p of C, then R is Zariski central.

<u>PROOF</u>. The center of $Q_{C-p}(R)$ is $Q_{C-p}(C)$, and $Q_{C-p}(R)$ is a central extension of $R/\kappa_{C-p}(R) = \overline{R}$. Let I be an ideal of $\overline{R}$ and put $J = Q_{C-p}(R)I$. Since κ_{C-p} is a t-functor,

$$Q_{C-p}(C) \cap J = Q_{C-p}(C)(J \cap \overline{C}).$$

By assumption we have :

$$J^n \subset Q_{C-p}(R)(J \cap Q_{C-p}(C)) = Q_{C-p}(R)(J \cap \overline{C}),$$

therefore $I^n \subset Q_{C-p}(R)(J \cap \overline{C})$ for some $n \geqslant 0$. Now since $\overline{C} \cap Q_{C-p}(R/I/\overline{C} \cap I)$ is κ_{C-p}-torsion, we have that :

$$Q_{C-p}(R)(\overline{C} \cap J) = Q_{C-p}(R)(\overline{C} \cap I).$$

Hence, $I^n \subset \overline{R} \cap Q_{C-p}(R)(\overline{C} \cap I)$, i.e. I^n is contained in the κ_{C-p}-closure of $\overline{R}(\overline{C} \cap I)$.

Now let I be an ideal of R and let j_p be the canonical ring morphism $j_p : R \to Q_{C-p}(R)$. The κ_{C-p}-closure of $R(I \cap C)$ maps onto the κ_{C-p}-closure of $j_p(R)(j_p(I) \cap j_p(C))$ hence I^n is contained in the κ_{C-p}-closure of $R(I \cap C)$ for some $n \geqslant 0$. Take n such that the foregoing holds for every maximal ideal p of C (there is only a finite number of such). If $x \in I^n$, then $A_p x \subset R(I \cap C)$, $A_p \in \mathcal{L}(\kappa_{C-p})$ for every p. Hence $(\sum_p A_p)x \subset R(I \cap C)$, but since $\sum_p A_p$ cannot be contained in any proper maximal ideal of C, we get $1 \in \sum_p A_p$ and $x \in R(I \cap C)$. Thus $I^n \subset R(I \cap C)$, proving that R is Zariski central. ■

CHAPTER II. REFLECTIONS ON REFLECTORS AND LOCALIZATION IN GROTHENDIECK CATEGORIES

II.1. Categorical Background

A category $\underline{C}$ is *preadditive* if each set $\mathrm{Hom}_{\underline{C}}(C,C')$ is an abelian group and the composition maps

$$\mathrm{Hom}_{\underline{C}}(C',C'') \times \mathrm{Hom}_{\underline{C}}(C,C') \to \mathrm{Hom}_{\underline{C}}(C,C''),$$

are bilinear. $\underline{C}$ is said to be *abelian* if :

A_1. $\underline{C}$ is preadditive.

A_2. Every finite family of objects has a product (and coproduct) in $\underline{C}$.

A_3. Every morphism has a kernel and a cokernel.

A_4. For every morphism f, the canonical morphism $\bar{f}$: Coker(ker f) $\to$ Ker(coker f) is an isomorphism.

EXAMPLE. R-mod and the category of abelian groups *Ab*, are abelian categories.
Suppose that C is an object of $\underline{C}$ having a family $(C_i)_{i \in I}$ of subobjects such that $\oplus_{i \in I} C_i$ exists.
Let $j : \oplus_{i \in I} C_i \to C$ be the canonical morphism, then Im j is called the *sum* of the objects C_i and it is denoted by $\sum_{i \in I} C_i$; if j is a monomorphism then the sum is direct and $\oplus_{i \in I} C_i \cong \sum_{i \in I} C_i$.
Dually, the kernel of the canonical morphism $p : C \to \prod_{i \in I} C/C_i$, is the *intersection* of the objects C_i; it is denoted by $\bigcap_{i \in I} C_i$.

A functor $T : \underline{C} \to \underline{D}$, between abelian categories is *additive* if for $f_1, f_2 \in \mathrm{Hom}_{\underline{C}}(C,C')$, $T(f_1 + f_2) = T(f_1) + T(f_2)$. T is *left exact* (*left semi-exact*) if each exact sequence :

$$0 \to C' \to C \to C'' \to 0$$

in $\underline{C}$ yields an exact sequence :

$$0 \to T(C') \to T(C) \to T(C''),$$

$$(\text{resp. } 0 \to T(C') \to T(C))$$

under the action of T. Dually one defines right (semi-) exact functors. If T is both left and right exact then it is said to be exact. In this case T maps arbitrary exact sequences into exact sequences. Let $j_1 : C_1 \to C$ and $j_2 : C_2 \to C$ be morphisms in $\underline{C}$. A pullback or fibred product for j_1, j_2 is an object P in $\underline{C}$ together with morphisms $p_1 : P \to C_1$, $p_2 : P \to C_2$ such that

1. $j_1 p_1 = j_2 p_2$
2. For every object Q in $\underline{C}$ and morphisms $q_1 : Q \to C_1$, $q_2 : Q \to C_2$ with $j_1 q_1 = j_2 q_2$, there is a unique morphism $g : Q \to P$ such that $p_1 g = q_1$ and $p_2 g = q_2$.

   The fibred product of C_1, C_2 over C will be denoted by $C_1 \times_C C_2$.

PROPOSITION II.1.1. With the above notations :

1. If j_i, $i = 1,2$, is a monomorphism (epimorphism), then so is p_{3-i}.
2. If j_i is the kernel of a morphism $f : C \to D$, then p_{3-i} is the kernel of $f \circ p_{3-i}$.

Note that the above proposition implies in particular that if j_1 is a monomorphism then we may view $C_1 \times_C C_2$ as the "inverse image" $j_2^{-1}(C_1)$, and if j_1 and j_2 are both monomorphisms, then $C_1 \times_C C_2$ may be interpreted as being the intersection of C_1 and C_2 in C.

Now let $\underline{C}$ be an abelian category. An object E of $\underline{C}$ is said to be <u>injective</u> if the functor $\mathrm{Hom}_{\underline{C}}(-,E) : \underline{C}^{opp} \to \underline{Ab}$, which is always left exact, is actually exact. This means that each exact diagram in $\underline{C}$:

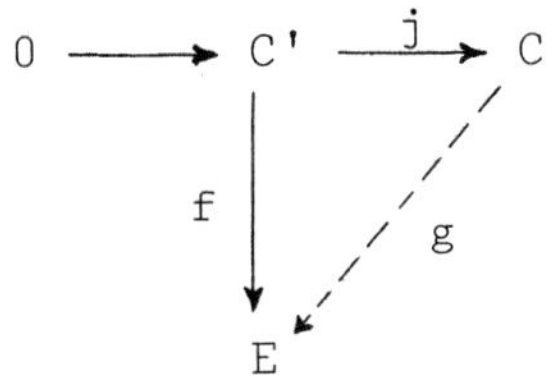

may be completed by a morphism $g : C \to E$ such that $gj = f$.

<u>PROPOSITION</u> II.1.2. 1. Let E be the direct sum of a family $(E_i)_{i \in I}$ of objects in $\underline{C}$. E is injective if and only if E_i is injective for every $i \in I$.

2. Let E be injective and $s : E \to M$ a monomorphism, then s is a section. Conversely, a section of an injective object is injective.

Recall that an additive functor $T : \underline{C} \to \underline{\mathcal{D}}$ between abelian categories is said to be <u>faithful</u> if $T(f) \neq 0$ for every non-zero morphism f in $\underline{C}$; equivalently : $T(C) \neq 0$ for every non-zero object C. An object G of $\underline{C}$ is a <u>generator</u> for $\underline{C}$ if $\mathrm{Hom}_{\underline{C}}(G,-)$ is faithful, and a <u>cogenerator</u> if $\mathrm{Hom}_{\underline{C}}(-,G)$ is faithful.

<u>PROPOSITION</u> II.1.3. Suppose $\underline{C}$ has coproducts. If G is a generator, then for each object C of $\underline{C}$ there is an epimorphism $G^{(I)} \to C$ for some index set I. Dually : suppose that $\underline{C}$ has products. If G is a cogenerator then for each object C of $\underline{C}$ there is a monomorphism $C \to G^I$ for some index set I.

EXAMPLE. The module $\mathrm{Hom}_{\mathbb{Z}}(R,\mathbb{Q}/\mathbb{Z})$ is an injective cogenerator for R-mod.

Recall furthermore that an abelian (or preadditive) category is said to be complete (resp. cocomplete) if projective limits (resp. inductive limits) exist for every functor $F : I \to \underline{C}$, where I is small, i.e. the class of objects of I is a set. A cocomplete abelian category $\underline{C}$ is called a Grothendieck category if inductive limits are exact in $\underline{C}$ and $\underline{C}$ has a generator.

PROPOSITION II.1.4. (cf. [13]). The following conditions are equivalent :

1. Inductive limits are exact in $\underline{C}$.
2. $\underline{C}$ satisfies A.5 : let $\{C_i\}_{i \in I}$ be a direct family of subobjects of C in $\underline{C}$, then for any subobject B of C we have : $(\sum_i C_i) \cap B = \sum_i (C_i \cap B)$.
3. For every morphism $f : B \to C$ and every direct family $(C_i)_{i \in I}$ of subobjects of C, we have the equality : $f^{-1}(\sum_i C_i) = \sum_i f^{-1}(C_i)$.

PROPOSITION II.1.5. Let $\underline{C}$ be a Grothendieck category with generator G. An object E of $\underline{C}$ is injective if and only if for every monomorphism $j : C \to G$ and every morphism $f : C \to E$, there is an $f' : G \to E$ such that $f'j = f$.

Let D be a subobject of C, represented by a monomorphism $m : D \to C$, then we say that m (or C) is an extension of D. An extension $m : D \to C$ is said to be essential if for every non-zero subobject $B \to C$ of C, the "intersection" $D \times_C B$ is non-zero, more precisely : the canonical morphism $D \times_C B \to D$ is not the zero-morphism.

If $f : A \to B$ and $g : B \to C$ are extensions, then $gf : A \to C$ is essential if and only if both f and g are essential. It is clear that the inductive limit of a filtering system of essential extensions of an object C of $\underline{C}$ is an essential extension of C.

<u>PROPOSITION</u> II.1.6. Let $m : D \to C$ be a monomorphism, then the following statements are equivalent :

1. m is an essential extension of D.
2. If $j : C \to B$ is a morphism such that jm is a monomorphism, then j is a monomorphism.

<u>PROPOSITION</u> II.1.7. Let $m : D \to C$ be a monomorphism in the Grothendieck category $\underline{C}$. There exists an epimorphism $p : C \to B$ such that pm is an essential extension of D.

<u>PROPOSITION</u> II.1.8. An object E of a Grothendieck category is exactly then injective when it has no proper essential extension in $\underline{C}$, i.e. every essential extension of E is an isomorphism.

<u>PROPOSITION</u> II.1.9. Let $m : D \to E$ be a monomorphism, E an injective object of $\underline{C}$. For every essential extension $j : D \to C$ there is a monomorphism $i : C \to E$ such that $ij = m$; in other words : an injective object containing D contains a copy of each essential extension of D.

Inspired by the foregoing results we define an <u>injective hull</u>

of an object C of $\underline{C}$ to be an essential extension $C \to E$, where E is injective. This amounts to saying that E is a maximal essential extension of C in $\underline{C}$. The injective hull of C is unique up to isomorphism in $\underline{C}$.

THEOREM II.1.10. Any object of a Grothendieck category has an injective hull.

II.2. Localization in Grothendieck Categories

This, and the following section, contain some of the basic results concerning localization in Grothendieck categories, most of them due to P. Gabriel. However, the Gabriel filter, which is most useful if one works in the category R-mod, seems to loose its meaning in the categories we aim to study later; therefore we focus especially on the concept of a kernel functor and consequently the build-up of localization theory here is a bit different from the presentation to be found in [9] or [7].

Throughout, $\underline{C}$ is a Grothendieck category with generator, hence with enough injectives.

A torsion theory for $\underline{C}$ is a pair of classes $(\mathcal{T},\mathcal{F})$ such that :

1. $\mathrm{Hom}_{\underline{C}}(T,F) = 0$ for all $T \in \mathcal{T}$, $F \in \mathcal{F}$.
2. If $\mathrm{Hom}_{\underline{C}}(C,F) = 0$ for all $F \in \mathcal{F}$, then $C \in \mathcal{T}$.
3. If $\mathrm{Hom}_{\underline{C}}(T,C) = 0$ for all $T \in \mathcal{T}$, then $C \in \mathcal{F}$.

$\mathcal{T}$ is called a torsion class and its objects are torsion objects, while $\mathcal{F}$ is a torsion-free class consisting of torsion-free objects. Any class $\mathcal{G}$ of objects of $\underline{C}$ generates a torsion theory as follows :

$$\mathcal{F} = \{F \in \underline{C},\ \mathrm{Hom}_{\underline{C}}(C,F) = 0 \ \text{ for all } C \in \mathcal{G}\}$$

$$\mathcal{T} = \{T \in \underline{C},\ \mathrm{Hom}_{\underline{C}}(T,F) = 0 \ \text{ for all } F \in \mathcal{F}\}.$$

The torsion theory $(\mathcal{T},\mathcal{F})$ is hereditary if and only if $\mathcal{T}$ is closed under subobjects.

PROPOSITION II.2.1. The following properties of a class $\mathcal{T}$ are equivalent :

1. $\mathcal{T}$ is a torsion class for some (hereditary) torsion theory.
2. $\mathcal{T}$ is closed under quotient objects, coproducts, extensions, and subobjects.

PROPOSITION II.2.2. The following properties of a class $\mathcal{F}$ are equivalent :

1. $\mathcal{F}$ is a torsion-free class for some (hereditary) torsion theory.
2. $\mathcal{F}$ is closed under subobjects, products, isomorphic copies (and injective hulls).

An (idempotent) radical in $\underline{C}$ is a subfunctor r of the identity in $\underline{C}$ satisfying $r(C/r(C)) = 0$ for every object C in $\underline{C}$ (and $rr = r$). An idempotent radical is said to be a kernel functor if it is left exact.

PROPOSITION II.2.3. There is a bijective correspondence between

1. Torsion theories for $\underline{C}$ and idempotent radicals in $\underline{C}$
2. Hereditary torsion theories and kernel functors.

In what follows we will restrict our attention to hereditary torsion theories, i.e. to kernel functors. If K is a kernel functor then the corresponding torsion theory, denoted by $(\mathcal{T}_K, \mathcal{F}_K)$ is given by $\mathcal{T}_K = \{C \in \underline{C},\ K(C) = C\}$, $\mathcal{F}_K = \{C \in \underline{C},\ K(C) = 0\}$.

<u>PROPOSITION</u> II.2.4. Let $0 \to P' \to P \to P'' \to 0$ be an exact sequence in $\underline{C}$ such that $P'' \in \mathcal{T}_K$ and $P \in \mathcal{F}_K$ for some kernel functor K in $\underline{C}$, then P is an essential extension of P' in $\underline{C}$.

<u>PROOF</u>. Let $Q \to P$ be any subobject of P, then $K(P) = 0$ yields $K(Q) = 0$. Let Q' be the kernel functor of the composed morphism $Q \to P \to P''$; then we have the following exact diagram in $\underline{C}$:

$$\begin{array}{ccccc}
 & & 0 & & 0 \\
 & & \downarrow & & \downarrow \\
0 & \to & Q' & \longrightarrow & P' \\
 & & \downarrow & & \downarrow \\
0 & \to & Q & \longrightarrow & P \\
 & & \downarrow & & \downarrow \\
 & & P'' & \xrightarrow{\cong} & P'' \\
 & & & & \downarrow \\
 & & & & 0
\end{array}$$

Therefore we get a (unique) morphism $Q' \to Q \times_P P'$. Suppose $Q \times_P P' = 0$, then $Q' = 0$ and $Q \to P''$ is a monomorphism, but then $K(Q) = Q$ and this contradicts $K(Q) = 0$, unless $Q = 0$. So if $Q \neq 0$ then $Q \times_P P' \neq 0$. ∎

Let K be a kernel functor in $\underline{C}$; then an object E of $\underline{C}$ is said to be <u>K-injective</u> if every exact diagram :

$$\begin{array}{ccccccccc} 0 & \to & C' & \xrightarrow{i} & C & \to & C'' & \to & 0 \\ & & \downarrow f & \swarrow g & & & & & \\ & & E & & & & & & \end{array}$$

with $C'' \in \mathcal{T}_K$, may be completed by a morphism $g : C \to E$, such that $gi = f$. If g is unique as such then E is said to be faithfully K-injective.

PROPOSITION II.2.5. The following statements are equivalent :

1. E is K-injective and K-torsion free
2. E is faithfully K-injective.

PROOF. 1. ⇒ 2. Consider the following exact diagram in $\mathcal{C}$:

$$\begin{array}{ccccccccc} 0 & \to & C' & \xrightarrow{i} & C & \xrightarrow{p} & C'' & \to & 0 \\ & & \downarrow f & \swarrow g & & & & & \\ & & E & & & & & & \end{array}$$

where $C'' \in \mathcal{T}_K$. Since E is K-injective there exists at least one morphism $g : C \to E$ such that $gi = f$. Suppose g_1 and g_2 have this property, then $(g_1 - g_2)i = 0$, hence $g_1 - g_2$ factorizes through C", i.e. there is a morphism $h : C'' \to E$ such that $g_1 - g_2 = hp$. Since $C'' \in \mathcal{T}_K$ and $E \in \mathcal{F}_K$, $h = 0$ hence $g_1 = g_2$.

2. ⇒ 1. Consider the following diagram :

$$\begin{array}{ccccccc} 0 & \to & K(E) & \to & K(E) & \to & 0 \\ & \overline{0} \searrow & \downarrow & & & & \\ & & E & & & & \end{array}$$

where $\overline{0}$ is the zero-map. Since $K(E) \in \mathcal{T}_K$ there is a unique extension of $\overline{0}$ to K(E), so it has to be the zero map too. Since K(E) $\to$ E is a monomorphism, K(E) = 0 follows. ■

The Propositions I.1.1., I.1.2., I.1.3. will be special cases of the foregoing and the following propositions if one takes $\underline{C}$ = R-mod.

PROPOSITION II.2.6. Let K be a kernel functor in $\underline{C}$, let $0 \to E' \xrightarrow{i} E \xrightarrow{p} E'' \to 0$ be an exact sequence in $\underline{C}$ such that E is K-injective and E" is K-torsion free, then E' is K-injective.

PROOF. Consider the following diagram wit exact rows and given morphism f' :

$$\begin{array}{ccccccccc}
0 & \to & E' & \xrightarrow{i} & E & \xrightarrow{p} & E'' & \to & 0 \\
 & & {\scriptstyle f'}\uparrow & & {\scriptstyle f}\uparrow & & {\scriptstyle f''}\uparrow & & \\
0 & \to & C' & \xrightarrow{j} & C & \to & C'' & \to & 0
\end{array}$$

where $C'' \in \mathcal{T}_K$. First, f is obtained using the K-injectivity of E and f" is the induced quotient ring. Since $E'' \in \mathcal{F}_K$ and $C'' \in \mathcal{T}_K$, it follows that f" = 0, hence f factorizes through E' : $f = if_1$ where $f_1 : C \to E'$. It is readily checked that $f_1 j = f'$, hence E' is K-injective.

PROPOSITION II.2.7. Let K be a kernel functor in $\underline{C}$ and let $0 \to E' \xrightarrow{i} E \xrightarrow{p} E'' \to 0$ be an exact sequence in $\underline{C}$, where E' is K-injective, $E'' \in \mathcal{T}_K$ and with $E \in \mathcal{F}_K$, then E is isomorphic to E'.

PROOF. By Proposition II.2.4., i is an essential morphism.

Since $E'' \in \mathcal{T}_K$, the following diagram :

$$\begin{array}{ccccccccc} 0 & \to & E' & \xrightarrow{i} & E & \xrightarrow{p} & E'' & \to & 0 \\ & & {\scriptstyle 1_{E'}}\downarrow & \swarrow {\scriptstyle j} & & & & & \\ & & E' & & & & & & \end{array}$$

may be completed, in a unique way, by $j : E \to E'$ such that $ji = 1_{E'}$. From Proposition II.1.6. it follows that j is a monomorphism, hence $ji = 1_{E'}$ implies $E \cong E'$. ■

Let K be a kernel functor in $\underline{C}$; the class of all faithfully K-injective objects in $\underline{C}$ forms a full subcategory of $\underline{C}$. This category will usually be denoted by $\underline{C}(K)$ and it is called the quotient category of $\underline{C}$ with respect to K. The canonical inclusion is denoted by $i_K : \underline{C}(K) \to \underline{C}$. Let C be in $\mathcal{F}_K$, then a K-injective hull of C is defined to be an essential extension $C \to E$ such that E is K-injective and $E/C \in \mathcal{T}_K$. Clearly any K-injective hull is in $\underline{C}(K)$.

PROPOSITION II.2.8. Every $C \in \mathcal{F}_K$ has an essentially unique K-injective hull.

PROOF. Let E be an injective hull of C in $\underline{C}$, then $E \in \mathcal{F}_K$. Consider the exact sequence

$$0 \to C \to E \to E/C \to 0$$

and define $E' = E \times_{E/C} K(E/C)$. By Proposition II.1.1. we may view E' as a subobject of E, hence $E' \in \mathcal{F}_K$. Moreover E/E' is then isomorphic to $(E/C)/K(E/C)$, hence $K(E/E') = 0$. Applying Proposition II.2.6. to the exact sequence

$$0 \to E' \to E \to E/E' \to 0,$$

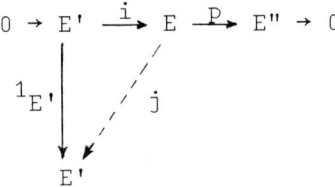

one obtains that E' is K-injective. On the other hand $E'/C \cong K(E/C)$ hence E'/C is K-torsion. Now suppose that E_1', E_2' are K-injective hulls of C, then E_2' is isomorphic to a subobject E_2'' of E_1' which contains C as a subobject. Because E_1' is in F_K and an essential extension of E_2'' which itself is faithfully K-injective, we may apply Proposition II.2.7. and conclude $E_1' \cong E_2'' \cong E_2'$.

The K-injective hull of $C \in \underline{C}$ will be denoted by $E_K(C)$. From the foregoing it follows that :

PROPOSITION II.2.9. $E_K(C)$ is the essentially unique faithfully K-injective object in $\underline{C}$ such that $E_K(C)/C$ is K-torsion.

If $\underline{C}_1$ and $\underline{C}_2$ are preadditive categories and $F : \underline{C}_1 \to \underline{C}_2$, $G : \underline{C}_2 \to \underline{C}_1$ additive functors then G is said to be a right adjoint of F (F is a left adjoint of G) if there is a natural equivalence Φ,

$$\Phi : \mathrm{Hom}_{\underline{C}_1}(-,G(-)) \to \mathrm{Hom}_{\underline{C}_2}(F(-),-)$$

of functors $\underline{C}_1^{opp} \times \underline{C}_2 \to \underline{Ab}$, i.e. for each pair of objects $C_1 \in \underline{C}_1$, $C_2 \in \underline{C}_2$ there is an isomorphism of abelian groups $\Phi_{C_1,C_2} : \mathrm{Hom}_{\underline{C}_1}(C_1,G(C_2)) \to \mathrm{Hom}_{\underline{C}_2}(F(C_1),C_2)$, which is natural in C_1 and C_2.

If G_1 and G_2 are right adjoints of F then G_1 and G_2 are naturally equivalent functors. A right adjoint preserves projective limits while a left adjoint preserves inductive limits. In a Grothendieck category $\underline{C}$, the following are equivalent for an endofunctor F :

1. F has a right adjoint
2. F is right exact and commutes with coproducts.

THEOREM II.2.10. *The inclusion functor* $i_K : \underline{C}(K) \to \underline{C}$ *has a left adjoint.*

PROOF. For $C \in \underline{C}$ define $\underline{\underline{a}}_K(C) = E_K(C/K(C))$, this definition yields a functor $\underline{\underline{a}}_K : \underline{C} \to \underline{C}(K)$. Let $f : C \to i_K(D)$ be an arbitrary morphism, where $C \in \underline{C}$, $D \in \underline{C}(K)$. Since $i_K(D) \in \mathcal{F}_K$, f extends to a morphism $f_1 : C/K(C) \to i_K(D)$. Since $\underline{\underline{a}}_K i_K(D)$ is faithfully K-injective and $\underline{\underline{a}}_K(C)/(C/K(C)) \in \mathcal{T}_K$ it follows that f_1 extends to $f' : \underline{\underline{a}}_K(C) \to \underline{\underline{a}}_K i_K(D) = D$. Finally, it is easily verified that we obtain an isomorphism :

$$\mathrm{Hom}_{\underline{C}}(C, i_K(D)) \cong \mathrm{Hom}_{\underline{C}(K)}(\underline{\underline{a}}_K(C), D). \qquad \blacksquare$$

Put $Q_K = i_K \underline{\underline{a}}_K$. For $C \in \underline{C}$, $Q_K(C)$, together with the canonical morphism $j_K : C \to Q_K(C)$ is called the *$\underline{C}$-object of quotients of C at K*.

PROPOSITION II.2.11. Q_K is a left exact endofunctor in $\underline{C}$.

PROOF. This follows from some general abstract nonsense but we prefer to give a direct proof here. If $0 \to C' \xrightarrow{i} C$ is exact in $\underline{C}$ then so is $0 \to C'/K(C') \to C/K(C)$. Now, since $Q_K(C')$, $Q_K(C)$ are essential extensions of $C'/K(C')$, $C/K(C)$ resp. it is not hard to see that $Q_K(i)$ is a monomorphism. First, let $C \in \mathcal{F}_K$ and consider the following commutative diagram with exact top row :

$$\begin{array}{ccccccc}
0 \to & C' & \xrightarrow{f} & C & \xrightarrow{g} & C'' \\
 & \downarrow & & \downarrow & & \downarrow \\
0 \to & Q_K(C') & \xrightarrow[Q_K(f)]{} & Q_K(C) & \xrightarrow[Q_K(g)]{} & Q_K(C'')
\end{array}$$

We know that $Q_K(f)$ is a monomorphism and that

$Q_K(g)Q_K(f) = Q_K(gf) = \overline{0}$. Hence $Q_K(C')$ is a subobject of Ker $Q_K(g)$. Consider the exact sequence :

$$0 \to \text{Ker } Q_K(g) \to Q_K(C) \to \text{Im } Q_K(g) \to 0,$$

since $Q_K(C)$ is K-injective, Im $Q_K(g)$ is K-torsion free, we may conclude that Ker $Q_K(g)$ is K-injective, hence faithfully K-injective. Moreover Ker $Q_K(g)/C' \cong Q_K(C)/C$ is K-torsion, hence Proposition II.2.9. yields that $Q_K(C') = \text{Ker } Q_K(g)$. In general, i.e. if C is not necessarily in F_K, consider

$$0 \to C' \xrightarrow{f} C \to C'' \to 0$$

and define $D = C \times_{C''} K(C'')$, then K(C) is clearly a subobject of D in C. By exactness of the sequence, Im f is a subobject of D and D/Im f ≅ K(C"). Therefore D/K(C) contains Im f + K(C)/K(C) and modulo the latter subobject is K-torsion. We have an exact sequence :

$$0 \to D/K(C) \to C/K(C) \to C''/K(C''),$$

where K(C/K(C)) = 0, hence we are reduced to proving the statement in the torsion free case. Indeed, we obtain an exact sequence :

$$0 \to Q_K(D/C) \to Q_K(C) \to Q_K(C''),$$

where

$$Q_K(D/C) = Q_K(\text{Im } f + K(C)/K(C)) = \text{Im } Q_K(f). \quad \blacksquare$$

Note that, although $\underline{\underline{a}}_K$ has right adjoint i_K, Q_K need not have a right adjoint or Q_K need not even be right exact.

II.3. Giraud Subcategories of a Complete Grothendieck Category

This section prepares for the study of Sheaves as a subcategory of Presheaves. $\mathcal{P}$ will denote a complete Grothendieck category. A full subcategory $\mathcal{S}$ of $\mathcal{P}$ is called reflective if the inclusion functor $i : \mathcal{S} \to \mathcal{P}$ has a left adjoint $\underline{\underline{a}}$, which will be called the reflector of $\mathcal{S}$ in $\mathcal{P}$, i.e. for $P \in \mathcal{P}$, $S \in \mathcal{S}$ there is a natural isomorphism $\mathrm{Hom}_{\mathcal{P}}(P,iS) \cong \mathrm{Hom}_{\mathcal{S}}(\underline{\underline{a}}P,S)$ with canonical natural transformations $p : \underline{\underline{a}}i \to 1_{\mathcal{S}}$ and $q : 1_{\mathcal{P}} \to i\underline{\underline{a}}$. The couple $(aP, q_P : P \to i\underline{\underline{a}}P)$ is universal in the following sense : every $\mathcal{P}$-morphism $f : P \to iS$ with $S \in \mathcal{S}$ can, in a unique way, be factorized through $i\underline{\underline{a}}P$ via q_P :

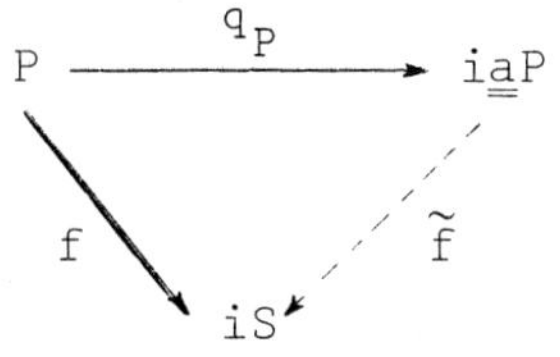

The morphism q_P is called the reflection of P.

PROPOSITION II.3.1. If $\mathcal{S}$ is a reflective subcategory of a complete Grothendieck category $\mathcal{P}$ then $\mathcal{S}$ is complete and cocomplete. If the reflector of $\mathcal{S}$ in $\mathcal{P}$ is left exact then $\mathcal{S}$ has exact inductive limits and a generator.

A subcategory of $\mathcal{P}$ with left exact reflector is called a Giraud subcategory of $\mathcal{P}$. From the definition it follows that, if $\mathcal{S}$ is a Giraud subcategory of a complete Grothendieck category $\mathcal{P}$, then $\mathcal{S}$ is a complete Grothendieck category and the reflector is exact, whereas the inclusion of $\mathcal{S}$ in $\mathcal{P}$ is generally only left exact.

Denote by $\mathcal{T}$ the class of objects C in $\mathcal{P}$ for which $\underline{\underline{a}}(C) = 0$ and let $\mathcal{F}$ consist of the subobjects in $\mathcal{P}$ of objects of $\mathcal{S}$.

<u>PROPOSITION</u> II.3.2. An object P of $\mathcal{P}$ is in $\mathcal{F}$ exactly then when the reflector $q_P : P \to i\underline{\underline{a}}P$ is a monomorphism.

<u>PROOF</u>. If q_P is a monomorphism then P is in $\mathcal{F}$ since $\underline{\underline{a}}P \in \mathcal{S}$. Conversely, let $P \in \mathcal{F}$ and $0 \to P \to iS$ with $S \in \mathcal{S}$. We have a commutative diagram in $\mathcal{P}$:

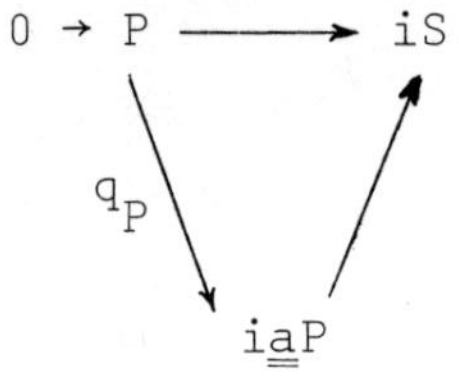

therefore q_P is a monomorphism. ■

$\mathcal{F}$ may be considered as a full complete subcategory of $\mathcal{P}$ which is easily verified to be a reflective subcategory of $\mathcal{P}$ with (epimorphic) reflector $\underline{\underline{a}}k$, where $\underline{\underline{a}}$ is the reflector of $\mathcal{S}$ and $k : \mathcal{F} \to \mathcal{P}$ the canonical inclusion. The objects of $\mathcal{F}$ are said to be <u>separated</u> (when there is no ambiguity concerning $\mathcal{S}$ or $\underline{\underline{a}}$).

<u>PROPOSITION</u> II.3.3. If $P \in \mathcal{P}$ is separated, then $i\underline{\underline{a}}P$ is an essential extension of P in $\mathcal{P}$.

<u>PROOF</u>. Let Q be a subobject of $i\underline{\underline{a}}P$ in $\mathcal{P}$ and assume that $Q \times_{i\underline{\underline{a}}P} P = 0$. Then $0 = \underline{\underline{a}}(Q \times_{i\underline{\underline{a}}P} P) = \underline{\underline{a}}\, Q \times_{\underline{\underline{a}}P} \underline{\underline{a}}P \cong \underline{\underline{a}}Q$. However, since Q is a subobject of a separated object, it is separated itself i.e. the canonical morphism $Q \to i\underline{\underline{a}}Q$ is a morphism, whence $Q = 0$ follows. ■

<u>PROPOSITION</u> II.3.4. Let $0 \to P' \to P \to P'' \to 0$ be an exact sequence in $\mathcal{P}$, then :

1. If $P' \in \mathcal{S}$, $P \in \mathcal{F}$, then $P'' \in \mathcal{F}$.
2. If $P'' \in \mathcal{F}$, $P \in \mathcal{S}$, then $P' \in \mathcal{S}$.

PROOF. Consider the following exact commutative diagram :

$$
\begin{array}{ccccccccc}
 & & 0 & & 0 & & 0 & & \\
 & & \downarrow & & \downarrow & & \downarrow & & \\
0 & \longrightarrow & \operatorname{Ker} q_{P'} & \longrightarrow & \operatorname{Ker} q_P & \longrightarrow & \operatorname{Ker} q_{P''} & & \\
 & & \downarrow & & \downarrow & & \downarrow & & \\
0 & \longrightarrow & P' & \longrightarrow & P & \longrightarrow & P'' & \longrightarrow & 0 \\
 & & q_{P'} \downarrow & & q_P \downarrow & & q_{P''} \downarrow & & \\
0 & \longrightarrow & i\underline{\underline{a}}P' & \longrightarrow & i\underline{\underline{a}}P & \longrightarrow & i\underline{\underline{a}}P'' & & \\
 & & \downarrow & & \downarrow & & \downarrow & & \\
 & \dashrightarrow & \operatorname{Coker} q_{P'} & \longrightarrow & \operatorname{Coker} q_P & \longrightarrow & \operatorname{Coker} q_{P''} & & \\
 & & \downarrow & & \downarrow & & \downarrow & & \\
 & & 0 & & 0 & & 0 & &
\end{array}
$$

In case 1. the snake lemma provides us with an exact sequence

$$0 \to 0^{(1)} \to 0^{(2)} \to \operatorname{Ker} q_{P''} \to 0^{(3)} \to \operatorname{coker} q_P \to \operatorname{coker} q_{P''}$$

(1) because $P' \in \mathcal{S}$, $\operatorname{Ker} q_{P'} = 0$

(2) because $P \in \mathcal{F}$, $\operatorname{Ker} q_P = 0$

(3) because $P' \in \mathcal{S}$, $\operatorname{Coker} q_{P'} = 0$.

Hence $\operatorname{Ker} q_{P''} = 0$, i.e. $P'' \in \mathcal{F}$.

In case 2. we get an exact sequence :

$$0 \to \operatorname{Ker} q_{P'} \to 0^{(1)} \to 0^{(2)} \to \operatorname{Coker} q_{P'} \to 0^{(3)} \to \operatorname{Coker} q_{P''}$$

(1) because $P \in \mathcal{S}$, $\operatorname{Ker} q_P = 0$

(2) because $P'' \in \mathcal{F}$, $\operatorname{Ker} q_{P''} = 0$

(3) because $P \in \mathcal{S}$, $\operatorname{Coker} q_P = 0$.

Hence $\operatorname{Ker} q_{P'} = \operatorname{Coker} q_{P'} = 0$, i.e. $P' \cong i\underline{\underline{a}}P'$ or $P' \in \mathcal{S}$. ∎

The foregoing proposition may also be derived from the following :

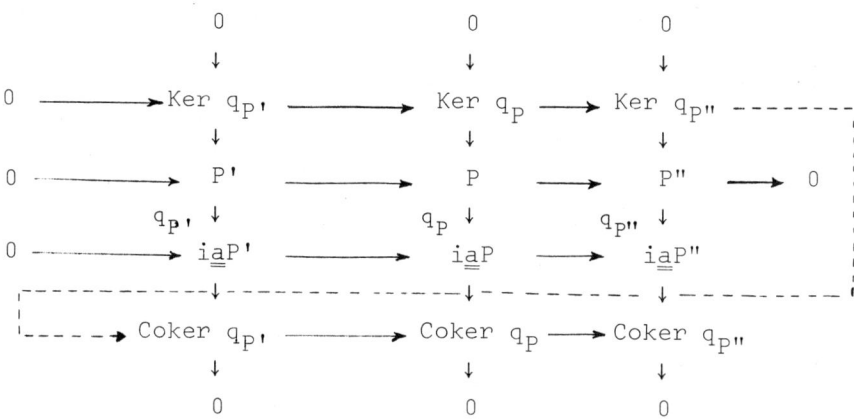

PROPOSITION II.3.5. The couple $(\mathcal{T},\mathcal{F})$ determines a torsion theory in $\mathcal{P}$ such that its quotient category is equivalent to $\mathcal{S}$.

PROOF. Since $\underline{\underline{a}}$ is exact, $\mathcal{T}$ is closed under subobjects, quotient objects and extensions. Since $\underline{\underline{a}}$ has a right adjoint it preserves coproducts, hence $\mathcal{T}$ may be considered as a torsion class. Obviously for $T \in \mathcal{T}$, $S \in \mathcal{S}$ we have $\mathrm{Hom}_{\mathcal{P}}(T,iS) = 0$ and also $\mathrm{Hom}_{\mathcal{P}}(T,F) = 0$ for $F \in \mathcal{F}$. Conversely, if $\mathrm{Hom}_{\mathcal{P}}(T,P) = 0$ for all $T \in \mathcal{T}$, then since $\mathrm{Ker}\ q_P \in \mathcal{T}$, we get that $\mathrm{Ker}\ q_P = 0$, hence $P \in \mathcal{F}$, therefore $\mathcal{F}$ can be considered as the torsion free class corresponding to $\mathcal{T}$. ■

The kernel functor associated to $(\mathcal{T},\mathcal{F})$ will be denoted by A. Clearly, if $P \in \mathcal{P}$, then :

$$A(P) = \sum\{P',\ 0 \to P' \to P \text{ and } \underline{\underline{a}}P' = 0\}.$$

A Giraud subcategory of $\mathcal{P}$ is said to be <u>strict</u> if it is closed under $\mathcal{P}$-isomorphisms.

PROPOSITION II.3.6. If K is a kernel functor in a complete Grothendieck category $\mathcal{P}$ then $\mathcal{P}(K)$ is a strict Giraud subcatecory of $\mathcal{P}$.

PROOF. Immediate consequence of Theorem II.2.10. and Proposition II.2.11. ■

We include some classical results of P. Gabriel and Gabriel-Popescu.

THEOREM II.3.7. (Gabriel). There is a bijective correspondence between torsion theories for $\mathcal{P}$ and strict Giraud subcategories of $\mathcal{P}$.

THEOREM II.3.8. (Gabriel-Popescu). Let $\underline{C}$ be a Grothendieck category with a generator G. Put $R = \mathrm{Hom}_{\underline{C}}(G,G)$ and let $T : \underline{C} \to R\text{-}\underline{\mathrm{mod}}$ be the functor $T(C) = \mathrm{Hom}_{\underline{C}}(G,C)$, $C \in \underline{C}$. Then :

1. T is full and faithfull
2. T induces an equivalence between $\underline{C}$ and $R\text{-}\underline{\mathrm{mod}}(K)$, where K is is the largest kernel functor in $R\text{-}\underline{\mathrm{mod}}$ for which all modules T(C) are faithfully K-injective.

COROLLARIES. Every object in a Grothendieck category has an injective extension. Every Grothendieck category is complete.

II.4. Kernel Functors in Giraud Subcategories

This section contains the fundamental results relating localization in Giraud subcategories to localization in the Grothendieck category under consideration.

Throughout this section $\mathcal{P}$ is a (complete) Grothendieck category, $\mathcal{S}$ is a (strict) Giraud subcategory of $\mathcal{P}$ which inclusion functor i and reflector $\underline{\underline{a}}$.

PROPOSITION II.4.1. An object E of $\mathcal{S}$ is injective in $\mathcal{S}$ if and only if iE is injective in $\mathcal{P}$.

PROOF. If iE is injective in $\mathcal{P}$, $E \in \mathcal{S}$, then E is injective in $\mathcal{S}$ because i is left exact. Conversely, let E be injective in $\mathcal{S}$ and consider a diagram in $\mathcal{P}$:

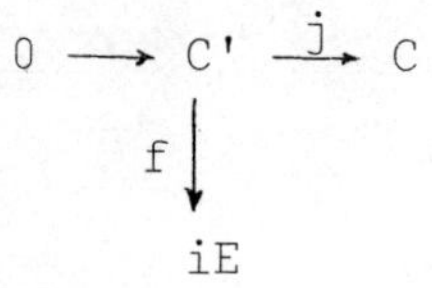

This diagram yields a commutative diagram in $\mathcal{P}$:

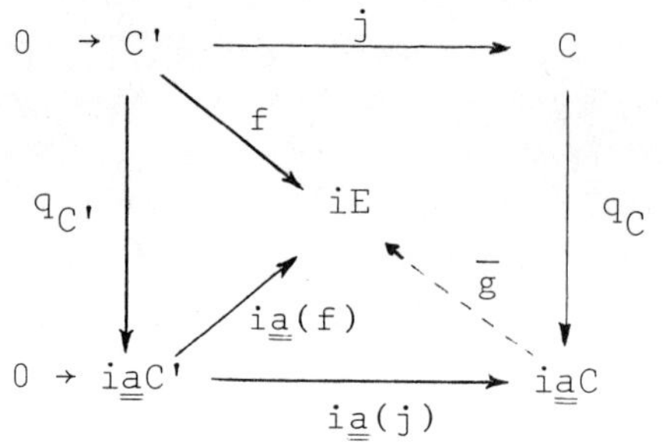

where the existence of $\overline{g}$ follows from the fact that E is injective in $\mathcal{S}$: $i\underline{\underline{a}}(f) = \overline{g} \circ i\underline{\underline{a}}(j)$. Put $g = \overline{g}\, q_C$, then $gj = \overline{g}\, q_C j = \overline{g}\, i\underline{\underline{a}}(j) q_{C'} = i\underline{\underline{a}}(f) q_{C'} = f$, what completes the proof. ■

If $C \in \mathcal{S}$, resp. $C \in \mathcal{P}$, then we will denote the injective hull of C in $\mathcal{S}$, resp. $\mathcal{P}$, by $E^S(C)$, resp. $E^P(C)$. The following lemmas and propositions relate $E^S(C)$ and $E^P(C)$.

<u>LEMMA</u> II.4.2. Let $S \in \mathcal{S}$, then $iE^S(S)$ is an essential extension of iS in $\mathcal{P}$.

<u>PROOF</u>. Let P be a subobject of $iE^S(S)$ in $\mathcal{P}$ and suppose that $P \underset{iE^S(S)}{\times} iS = 0$. Exactness of $\underline{\underline{a}}$ yields :

$$0 = \underline{\underline{a}}(0) = \underline{\underline{a}}(P \underset{iE^S(S)}{\times} iS) = \underline{\underline{a}}P \underset{\underline{\underline{a}}iE^S(S)}{\times} \underline{\underline{a}}iS = \underline{\underline{a}}P \underset{E^S(S)}{\times} S$$

However, this contradicts the fact that $E^S(S)$ is an essential extension of S in $\mathcal{S}$, since $\underline{\underline{a}}P \in \mathcal{S}$. ■

LEMMA II.4.3. Let $S \in \mathcal{S}$ then $E^P(iS) = iE^S(S)$; i.e. the hull in P of an object in $\mathcal{S}$ is in $\mathcal{S}$ too.

PROOF. By the foregoing lemma, $iE^S(S)$ as well as $E^P(iS)$ is an essential extension of iS, therefore we obtain a commutative diagram in P :

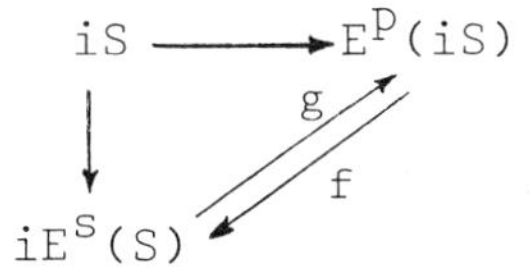

where f exists because of Proposition II.4.1. and it is a monomorphism, and where g is a monomorphism too. Now since $E^P(iS)$ is a maximal essential extension of iS in P it follows from $0 \to E^P(iS) \xrightarrow{f} iE^S(S)$, that $E^P(iS) \cong iE^S(S)$.

COROLLARY II.4.4. If $P \in \mathcal{P}$ is separated then $E^P(P)$ is separated. Indeed we have a commutative diagram of monomorphisms in P :

$$\begin{array}{ccc} P & \longrightarrow & \underline{ia}P \\ \downarrow & & \downarrow \\ E^P(P) & \longrightarrow & E^P(\underline{ia}P) \end{array}$$

Since $E^P(\underline{ia}P) \cong iE^S(\underline{a}P)$ it follows that $E^P(P)$, being a subobject in P of $E^S(\underline{a}P) \in \mathcal{S}$, is separated.

PROPOSITION II.4.5. If $P \in \mathcal{P}$ is separated then we have :

$$\underline{ia}E^P(P) \stackrel{(1)}{=} E^P(\underline{ia}P) \stackrel{(2)}{=} iE^S(\underline{a}P) \stackrel{(3)}{=} E^P(P).$$

PROOF. (1) It is clear that we have monomorphisms :

$$P \to \underline{ia}P; \quad E^P(P) \to E^P(\underline{ia}P); \quad \underline{ia}E^P(P) \to E^P(\underline{ia}P).$$

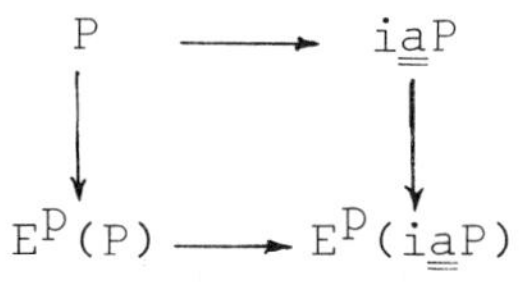

Now $E^P(i\underline{\underline{a}}P)$ is an essential extension of $i\underline{\underline{a}}P$ in $\mathcal{P}$ and $i\underline{\underline{a}}P$ is an essential extension of P in $\mathcal{P}$, then $E^P(i\underline{\underline{a}}P)$ is an essential extension in P of $\mathcal{P}$. To prove (1) it will therefore be sufficient to establish that $i\underline{\underline{a}}E^P(P)$ is injective in $\mathcal{P}$ or equivalently, that $\underline{\underline{a}}E^P(P)$ is injective in $\mathcal{S}$, cf. Proposition II.4.1.

Let $0 \to S' \xrightarrow{s} S$ be exact in $\mathcal{S}$ and let $f : S' \to \underline{\underline{a}}E^P(P)$ be a given $\mathcal{S}$-homomorphism. The definition of the fibred product gives rise to the following commutative diagram :

$$\begin{array}{ccccc} S_1' = iS' \underset{i\underline{\underline{a}}E^P(P)}{\times} E^P(P) & \xrightarrow{s'} & iS' & \xrightarrow{is} & iS \\ \downarrow f_1 & & \downarrow if & \swarrow ig & \\ E^P(P) & \longrightarrow & i\underline{\underline{a}}E^P(P) & & \end{array}$$

where s' is a monomorphism, hence (is)s' is a monomorphism. It is obvious that :

$$\underline{\underline{a}}S_1' = S' \underset{\underline{\underline{a}}E^P(P)}{\times} \underline{\underline{a}}E^P(P) \simeq S'.$$

Injectivity of $E^P(P)$ in $\mathcal{P}$ implies that there is a $\mathcal{P}$-morphism $g_1 : iS \to E^P(P)$, such that $g(is)s' = f_1$. Put $g = \underline{\underline{a}}(g_1) : S \to \underline{\underline{a}}E^P(P)$. Let j be the isomorphism $j : \underline{\underline{a}}S_1' \to S'$. Then,

$$\underline{\underline{a}}f_1 = f_j = q(g_1)s\,\underline{\underline{a}}(s') = gsj,$$

but fj = gsj yields f = gs since j is an isomorphism and in particular an epimorphism in $\mathcal{S}$.

(2) Lemma II.4.3.

(3) P is separated, so $E^P(P)$ is separated by the corollary to Lemma II.4.3. Then $i\underline{\underline{a}}E^P(P)$ is essential over $E^P(P)$ hence over P in $\mathcal{P}$, yielding $E^P(P) = i\underline{\underline{a}}E^P(P)$ because $E^P(P)$ is a maximal essential

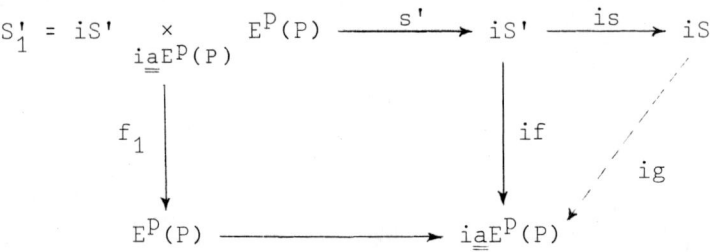

extension of P in $\mathcal{P}$. ■

Let K,K' be kernel functors in $\mathcal{P}$, then $K \geq K'$ if and only if $\mathcal{T}_K \supset \mathcal{T}_{K'}$ equivalently $K(P) \supset K'(P)$ for every $P \in \mathcal{P}$, this gives a partial ordering for kernel functors in $\mathcal{P}$.

Let K,K' be kernel functors in $\mathcal{P}$, then K' is said to be Q_K-compatible if $K'Q_K = Q_K K'$, where $i_K : \mathcal{P}(K) \to \mathcal{P}$ is the canonical inclusion.

LEMMA II.4.6. Let $P \in \mathcal{P}$, let K,K' be kernel functors in $\mathcal{P}$ such that K' is Q_K-compatible then :

1. If P is K'-torsion free, then $Q_K(P)$ is K'-torsion free; the converse holds if P is K-torsion free.
2. If P is K'-torsion, then $Q_K(P)$ is K'-torsion; the converse holds if P is K-torsion free.

PROOF. Easy. Note that this lemma states exactly that, for a Q_K-compatible kernel functor K', Q_K is inner in $\mathcal{T}_{K'}$ as well as in $\mathcal{F}_{K'}$. ■

To any strict Giraud subcategory $\mathcal{S}$ of $\mathcal{P}$ we may canonically associate a kernel functor A in $\mathcal{P}$. A-compatible kernel functors may then sometimes be referred to as $\mathcal{S}$-compatible(*) kernel functors. Hence a kernel functor K in $\mathcal{P}$ is $\mathcal{S}$-compatible if and only if $i\underline{\underline{a}}K = Ki\underline{\underline{a}}$. If $i\underline{\underline{a}}Ki\underline{\underline{a}} = Ki\underline{\underline{a}}$, i.e. K maps objects of $\mathcal{S}$ to objects

---

(*) Some authors (cf. [32], [33]) prefer " G-functor" , where G is used en hommage à Gabriel, Giraud, Golan, Goldie, Goldman, Grothendieck, G. Michler,...

of $\mathcal{S}$, then K is said to be <u>inner</u> in $\mathcal{S}$. If K is inner in $\mathcal{S}$, then the functor Ki will be denoted by K^S. It is in general not necessarily a kernel functor in $\mathcal{S}$, however :

<u>PROPOSITION</u> II.4.7. Let K be an $\mathcal{S}$-compatible kernel functor in $\mathcal{P}$, then K^S is a kernel functor in $\mathcal{S}$.

<u>PROOF</u>. If $S \in \mathcal{S}$, then $K(iS) = K(i\underline{\underline{a}}iS) = i\underline{\underline{a}}K(iS)$, hence K^S is inner in $\mathcal{S}$. Therefore $K^S(S) = \underline{\underline{a}}K(iS)$ and it is immediately clear that K^S is a left exact subfunctor of the identity in $\mathcal{S}$. Furthermore :

$$K^S(S/K^S(S)) = \underline{\underline{a}}K(ia(iS/iK^S(S))) =$$

$$= \underline{\underline{a}}K(iS/iK^S(S)) = \underline{\underline{a}}K(iS/K(iS)) = \underline{\underline{a}}0 = 0. \quad \blacksquare$$

<u>PROPOSITION</u> II.4.8. Let K,K'be kernel functors in $\mathcal{P}$, if $K \geq K'$, then K'is $\mathcal{P}(K')$-compatible.

<u>PROOF</u>. Let $P \in \mathcal{P}$ and let $q_{K(P)} : K(P) \to Q_{K'}(K(P))$ be the reflection of P in $\mathcal{P}(K')$.

We obtain the following exact sequence :

$$0 \to \mathrm{Im}\, q_{K(P)} \to Q_{K'}(K(P)) \to \mathrm{Coim}\, q_{K(P)} \to 0.$$

It is clear that Im $q_{K(P)}$ is K-torsion, moreover $K \geq K'$implies that Coim $q_{K(P)}$ is K-torsion, hence $Q_{K'}(K(P))$ is a subobject of $KQ_{K'}(P)$. Conversely, since $\mathcal{F}_K \subset \mathcal{F}_{K'}$ we have that $P/K(P) \in \mathcal{F}_{K'}$,hence

$$0 \to P/K(P) \to Q_{K'}(P/K(P))$$

and $KQ_{K'}(P/K(P)) \cap P/K(P) = 0$ implies then that $Q_{K'}(P/K(P)) = 0$. The exactness of

$$0 \to Q_{K'}(K(P)) \to Q_{K'}(P) \to Q_{K'}(P/K(P))$$

entails that $KQ_{K'}(P) \subset Q_{K'}(K(P))$. ■

THEOREM II.4.9. If S is a strict Giraud subcategory of P and K an S-compatible kernel functor in P then $S \in S$ is (faithfully) K^S-injective if and only if iS is (faithfully) K-injective.

PROOF. Let iS be K-injective and consider a diagram in S :

$$\begin{array}{l} 0 \to S_1 \to S_2 \to S_2/S_1 \to 0 \\ \quad\ \ \downarrow f \\ \quad\ \ S \end{array}$$

with $K^S(S_2/S_1) = S_2/S_1$. In P this yields a diagram

$$\begin{array}{l} 0 \to iS_1 \to iS_2 \to iS_2/iS_1 \to 0 \\ \quad\ \ \downarrow if \\ \quad\ \ iS \end{array}$$

where iS_2/iS_1 is a subobject of $i(S_2/S_1)$ and $S_2/S_1 = \underline{\underline{a}}(iS_2/iS_1)$. Since S_2/S_1 is K^S-torsion, it follows from Lemma II.4.6. that iS_2/iS_1 is K-torsion, so there is $g' : iS_2 \to iS$ completing the above diagram. Obviously, $g = \underline{\underline{a}}(g')$ completes the diagram in S. Conversely, let S be K^S-injective and consider the following exact sequence in P :

$$\begin{array}{l} 0 \to P_1 \to P_2 \to P_2/P_1 \to 0 \\ \quad f \downarrow \\ \quad\ \ iS \end{array}$$

where $P_2/P_1 \in T_K$. Since $\underline{\underline{a}}$ is exact we obtain a diagram in S :

$$0 \to \underline{\underline{a}}P_1 \to \underline{\underline{a}}P_2 \to \underline{\underline{a}}(P_2/P_1) \to 0$$
$$\underline{\underline{a}}f \downarrow$$
$$S$$

Lemma II.4.6. yields that $\underline{\underline{a}}(P_2/P_1)$ is K^S-torsion, therefore there exists an $\mathcal{S}$-morphism $g' : \underline{\underline{a}}P_2 \to S$ completing the diagram. Let g be $(ig')q_{P_2} : P_2 \to iS$, then it is easily verified that g extends f as desired. Since iS is K-torsion free if and only if S is K^S-torsion free (because iS is separated and by Lemma II.4.6.), Proposition II.2.5. finishes the proof. ■

PROPOSITION II.4.10. Let K be an $\mathcal{S}$-compatible kernel functor in $\mathcal{P}$ and let S be an object in $\mathcal{S}$ which is K^S-torsion free; then $iE_{K^S}(S) \cong E_K(iS)$.

PROOF. By Lemma II.4.6., $iS \in \mathcal{F}_K$. By the foregoing, the fact that $E_{K^S}(S)$ is faithfully K^S-injective implies that $iE_{K^S}(S)$ is faithfully K-injective. Furthermore, $iE_{K^S}(S)/iS$ is K-torsion in $\mathcal{P}$ since $E_{K^S}(S)/S$ is K^S-torsion in $\mathcal{S}$. But $E_K(iS)$ is unique up to isomorphism in $\mathcal{P}$ with the properties mentioned above, therefore $E_K(iS) \cong iE_{K^S}(S)$. ■

COROLLARY. If K is an $\mathcal{S}$-compatible kernel functor in $\mathcal{P}$, and let $S \in \mathcal{S}$, then $iQ_{K^S}(S) = Q_K(iS)$.

PROOF. If S is K^S-torsion free, then this is an immediate consequence of the foregoing proposition. In general :

$$iQ_K(S) = iE_{K^S}(S/K^S(S)) = E_K(i\underline{\underline{a}}(iS/K(iS))).$$

Now iS/K(iS) is separated, cf. Proposition II.3.4., and thus

$i\underline{\underline{a}}(iS/K(iS))$ is an essential extension in $\mathcal{P}$ of $iS/K(iS)$. Hence, $i(S/K^S(S))$ is K-torsion free and

$Q_K(iS) = E_K(iS/K(iS)) \cong E_K(i\underline{\underline{a}}(iS/K(iS)))$. ■

PROPOSITION II.4.11. Let $\mathcal{S}$ be a strict Giraud subcategory of $\mathcal{P}$ and let K be a kernel functor in $\mathcal{P}$, such that $K \geqslant A$, where A is the kernel functor in $\mathcal{P}$ corresponding to $\mathcal{S}$. Then $Q_K(P) \in i\mathcal{S}$ for every $P \in \mathcal{P}$.

PROOF. Put $\overline{P} = P/K(P)$. Since $K \geqslant A$, $A(\overline{P}) = 0$ hence $\overline{P}$ is separated with respect to $\mathcal{S}$. Moreover, $\underline{\underline{a}}(i\underline{\underline{a}}\overline{P}/\overline{P}) = 0$ entails that $i\underline{\underline{a}}\overline{P}/\overline{P}$ is A-torsion hence K-torsion. Since $Q_K(P)$ is faithfully K-injective we obtain the following commutative diagram in $\mathcal{P}$:

$$\begin{array}{ccccccccc} 0 & \rightarrow & \overline{P} & \rightarrow & i\underline{\underline{a}}\overline{P} & \rightarrow & i\underline{\underline{a}}\overline{P}/\overline{P} & \rightarrow & 0 \\ & & {\scriptstyle q_{\overline{P}}}\downarrow & \swarrow {\scriptstyle j_P} & & & & & \\ & & Q_K(P) & & & & & & \end{array}$$

Because $q_{\overline{P}}$ is a monomorphism and since $i\underline{\underline{a}}\overline{P}$ is an essential extension of $\overline{P}$ in $\mathcal{P}$, j_P is a monomorphism and thus $Q_K(i\underline{\underline{a}}\overline{P}) = Q_K(P)$. By Proposition II.4.8., K is $\mathcal{S}$-compatible, and by the corollary to Proposition II.4.10.,

$$iQ_{K^S}(\underline{\underline{a}}\overline{P}) = Q_K(i\underline{\underline{a}}\overline{P}) = Q_K(P).$$

Thus $Q_K(P) \in i\mathcal{S}$. ■

To end this section we construct kernel functors associated to objects of a strict Giraud subcategory $\mathcal{S}$; they are all $\mathcal{S}$-compatible. Let P be a non-zero object in $\mathcal{P}$. Let Kerf(P) be the class of kernel functors K in $\mathcal{P}$ such that $P \in \mathcal{F}_K$. If $P' \in \mathcal{P}$ is an essential extension of P in $\mathcal{P}$ then obviously Kerf(P') = Kerf(P).

Therefore, when describing Kerf(P), we may suppose that P equals its injective hull $E^{\mathcal{P}}(P)$ in $\mathcal{P}$. Define K_P by its action on objects Q in $\mathcal{P}$ as follows :

$$K_P(Q) = \cap\{\text{Ker } g,\ g \in \text{Hom}_{\mathcal{P}}(Q, E^{\mathcal{P}}(P))\}.$$

PROPOSITION II.4.12. K_P is a kernel functor in $\mathcal{P}$, $K_P \in$ Kerf(P) and if K is a kernel functor in $\mathcal{P}$ then K $\in$ Kerf(P) if and only if $K \leq K_P$.

PROOF. It is straightforward to check that K_P is a kernel functor in $\mathcal{P}$. Since there is a monomorphism $0 \to E^{\mathcal{P}}(P) \to E^{\mathcal{P}}(P)$, $K_P(E^{\mathcal{P}}(P)) = 0$, i.e. $K_P \in$ Kerf(P).
If $K \leq K_P$, then $K(E^{\mathcal{P}}(P)) = 0$ and K $\in$ Kerf(P) follows. Conversely, assume that $K(E^{\mathcal{P}}(P)) = 0$ for some kernel functor K in $\mathcal{P}$, and let $g : Q \to E^{\mathcal{P}}(P)$ be an arbitrary morphism in $\mathcal{P}$, then obviously $K(Q) \subset \text{Ker } g$, hence $K(Q) \subset K_P(Q)$ for every $Q \in \mathcal{P}$; thus $K \leq K_P$. ■

THEOREM II.4.13. If $S \in \mathcal{S}$ then K_{iS} is an $\mathcal{S}$-compatible kernel functor in $\mathcal{P}$. Conversely if $P \in \mathcal{P}$ is such that K_P is $\mathcal{S}$-compatible then there is an $S \in \mathcal{S}$ such that $K_P = K_{iS}$; moreover $S = \underline{\underline{a}}P$.

PROOF. First we have to check that $i\underline{\underline{a}}K_{iS}(P) = K_{iS}(i\underline{\underline{a}}P)$, $P \in \mathcal{P}$. We may assume S replaced by $E^{\mathcal{S}}(S)$ and iS replaced by $E^{\mathcal{P}}(iS)$. The problem is reduced to the proof of :

$$\cap\{\text{Ker } g,\ g \in \text{Hom}_{\mathcal{P}}(i\underline{\underline{a}}P, iS)\} = i\underline{\underline{a}}(\cap\{\text{Ker } g,\ g \in \text{Hom}_{\mathcal{P}}(P, iS)\}).$$

Since $\mathcal{S}$ is a full subcategory of $\mathcal{P}$ and since $\underline{\underline{a}}$ is right adjoint to i we have the following isomorphisms in Ab :

$$\text{Hom}_{\mathcal{P}}(i\underline{\underline{a}}P, iS) \cong \text{Hom}_{\mathcal{S}}(\underline{\underline{a}}P, S) \cong \text{Hom}_{\mathcal{P}}(P, iS).$$

Clearly, $\cap\{\text{Ker } g,\ g \in \text{Hom}_{\mathcal{S}}(\underline{\underline{a}}P, S)\}$ is an object of $\mathcal{S}$, $\mathcal{S}$ being a

Giraud subcategory of $\mathcal{P}$, so we are done. Conversely, let $P \in \mathcal{P}$ be such that K_P is $\mathcal{S}$-compatible, then :

$$K_P(i\underline{\underline{a}}P) = i\underline{\underline{a}}K_P(P) = i\underline{\underline{a}}0 = 0,$$

hence $K_P \leqslant K_{i\underline{\underline{a}}P}$. On the other hand :

$$K_{i\underline{\underline{a}}P}(P) = \cap\{\text{Ker } g,\ g \in \text{Hom}_{\mathcal{P}}(P, E^P(i\underline{\underline{a}}P))\}.$$

Applying Lemma II.4.3. we get :

$$K_{i\underline{\underline{a}}P}(P) = \cap\{\text{Ker } g,\ g \in \text{Hom}_{\mathcal{P}}(P, iE^S(\underline{\underline{a}}P))\},$$

thus

$$K_{i\underline{\underline{a}}P}(P) = \cap\{\text{Ker } g,\ g \in \text{Hom}_{\mathcal{P}}(\underline{\underline{a}}P, E^S(\underline{\underline{a}}P))\} = 0.$$

Now $0 = K_{i\underline{\underline{a}}P}(P)$ yields $K_{i\underline{\underline{a}}P} \leqslant K_P$. ■

II.5. The Main Example : Sheaves and Presheaves

Let X be a topological space and let $\underline{C}$ be a Grothendieck category. The functor category $\underline{\text{Hom}}(\text{Open }(X)^{\text{opp}}, \underline{C})$ is the category of presheaves over X with values in $\underline{C}$. It will be denoted by $\mathcal{P}(X, \underline{C})$. A presheaf P assigns to each open set U an object P(U) of $\underline{C}$ and to each inclusion $V \subset U$ of open sets a morphism $P_V^U : P(U) \to P(V)$. A morphism $f : P \to P'$ of presheaves is given by a family of morphisms $f(U) : P(U) \to P'(U)$, such that for each couple of open sets $U \subset V$ we have a commutative diagram in $\underline{C}$:

$$\begin{array}{ccc} P(V) & \xrightarrow{f(V)} & P'(V) \\ \downarrow P_U^V & & \downarrow P'^V_U \\ P(U) & \xrightarrow{f(U)} & P'(U) \end{array}$$

If $U \in \text{Open }(X)$, let $\text{Cov}_X(U)$ be the set of all families

$\mathcal{U} = \{U_i\}_{i \in I}$ with $U_i \in$ Open (X) for each $i \in I$ and $\cup\mathcal{U} = U$. Take $\mathcal{U} \in \mathrm{Cov}_X(U)$, $U \in$ open (X) and let $P \in \mathcal{P}(X,\underline{C})$

Let $p_i : \prod_{i \in I} P(U_i) \to P(U_i)$ be the projection morphism.

We have a morphism $j : P(U) \to \prod_{i \in I} P(U_i)$ such that $p_i j = P^U_{U_i}$ and we also have morphisms

$$p,q : \prod_{i \in I} P(U_i) \rightrightarrows \prod_{(j,k) \in I \times I} P(U_i \cap U_j)$$

where the (j,k)-component of p is $P^{U_j}_{U_j \cap U_k}(p_j)$ and the (j,k)-component of q is $P^{U_k}_{U_j \cap U_k}(p_k)$. We get a diagram :

$$P(U) \xrightarrow{j} \prod_{i \in I} P(U_i) \underset{q}{\overset{p}{\rightrightarrows}} \prod_{(j,k) \in I \times I} P(U_i \cap U_j).$$

P is a sheaf if and only if this diagram is an equalizer diagram. Let us denote by $\mathcal{S}(X,C)$ the full subcategory of $\mathcal{P}(X,C)$, consisting of all sheaves over X with values in $\underline{C}$. It is easily checked that $\mathcal{S}(X,\underline{C})$ and $\mathcal{P}(X,\underline{C})$ are complete.

<u>THEOREM</u> II.5.1. <u>Let</u> $\underline{C}$ <u>be a Grothendieck category</u>, <u>then</u> :

1. $\mathcal{P}(X,\underline{C})$ <u>is a Grothendieck category</u>.
2. $\mathcal{S}(X,\underline{C})$ <u>is a Giraud subcategory of</u> $\mathcal{P}(X,\underline{C})$
3. <u>A presheaf</u> P <u>is separated if and only if</u> j <u>is a monomorphism</u>.

<u>PROOF</u>. Well known, cf. [1], [13]. ■

We recall some basic facts. The construction of the reflector $\underline{\underline{a}}$ for $\mathcal{S}(X,\underline{C})$ is executed in two steps. First, define a functor $L : \mathcal{P}(X,\underline{C}) \to \mathcal{P}(X,\underline{C})$, as follows. Let U be open in X, we give $\mathrm{Cov}_X(U)$ the structure of a category : if $\mathcal{U} = \{U_i\}_{i \in I}$, $\mathcal{V} = \{V_j\}_{j \in J}$ are in $\mathrm{Cov}_X(U)$, a morphism $\mathcal{U} \to \mathcal{V}$ is given by a map $\varepsilon : I \to J$

such that $U_i \subset V_{\varepsilon(i)}$ for all $i \in I$. Let $P \in \mathcal{P}(X,\underline{C})$ and define $[P,U]$, $U \in \mathrm{Open}\ (X)$, by its action on a covering $\mathcal{U} = \{U_i\}_{i \in I}$ of U :

$$[P,U](\mathcal{U}) = \mathrm{Ker}\big(\prod_{i \in I} P(U_i) \underset{q}{\overset{p}{\rightrightarrows}} \prod_{(j,k) \in I \times I} P(U_j \cap U_k)\big).$$

Note that $[P,U] : \mathrm{Cov}_X(U) \to \underline{C}$ is a contravariant functor. Hence we can define an object LP of $\mathcal{P}(X,\underline{C})$ by :

$$LP : \mathrm{Open}\ (X)^{\mathrm{opp}} \to \underline{\mathrm{Ens}} : U \to \varinjlim_{\mathcal{U} \in \mathrm{Cov}_X(U)} [P,U](\mathcal{U}).$$

Note that

$$LP(U) = \varinjlim_{\mathcal{U} \in \mathrm{Cov}_X(U)} \varprojlim_{V \in \mathcal{U}} P(V).$$

The assignment $P \to LP$ defines a left exact endofunctor of $\mathcal{P}(X,\underline{C})$.

The following lemmas are well known and straightforward to prove.

<u>LEMMA</u> II.5.2. Let $\mathcal{F}(X,\underline{C})$ be the class of separated objects in $\mathcal{P}(X,\underline{C})$, then :

1. If $P \in \mathcal{F}(X,\underline{C})$, then the canonical morphism $P \to LP$ is a monomorphism and $LP \in \mathcal{S}(X,\underline{C})$.
2. If $P \in \mathcal{P}(X,\underline{C})$, then $LP \in \mathcal{F}(X,C)$.
3. If $P \in \mathcal{S}(X,\underline{C})$, then $LP \cong P$ and conversely.

Define $i\underline{\underline{a}} = LL$, where $i : \mathcal{S}(X,\underline{C}) \hookrightarrow \mathcal{P}(X,\underline{C})$ is the canonical inclusion, then

<u>LEMMA</u> II.5.3. $\underline{\underline{a}}$ is a left adjoint of i.

Theorem II.5.1. allows to read $\mathcal{P}(X,\underline{C}),\mathcal{S}(X,\underline{C})$ for $\mathcal{P}$, $\mathcal{S}$ in Section II.3. and II.4.

Let us now look at the behaviour of sheaves and presheaves under change of underlying topological spaces. Let X,Y be topological spaces and let $f : X \to Y$ be a continuous map. To a presheaf $P \in \mathcal{P}(X,\underline{C})$ we associate a presheaf $f_* P \in \mathcal{P}(Y,\underline{C})$, given by putting $f_*P(U) = P(f^{-1}(U))$. This yields a functor $f_* : \mathcal{P}(X,\underline{C}) \to \mathcal{P}(Y,\underline{C})$ which restricts to a functor $\mathcal{S}(X,\underline{C}) \to \mathcal{S}(Y,\underline{C})$. Note that f_* is exact on $\mathcal{P}(X,\underline{C})$ but it is only left exact on $\mathcal{S}(X,\underline{C})$. If Z is a topological space, $g : Y \to Z$ a continuous map then $(gf)_* = g_*f_*$. The functor $f_* : \mathcal{P}(X,\underline{C}) \to \mathcal{P}(Y,\underline{C})$ has a left adjoint $f^\circ : \mathcal{P}(Y,\underline{C}) \to \mathcal{P}(X,\underline{C})$, defined by $f^\circ P(U) = \varinjlim_{f(U) \subset V} P(V)$. If $i_Y : \mathcal{S}(Y,\underline{C}) \to \mathcal{P}(Y,\underline{C})$ is the canonical inclusion and $\underline{\underline{a}}_X : \mathcal{P}(X,\underline{C}) \to \mathcal{S}(X,\underline{C})$ the reflector for $\mathcal{S}(X,\underline{C})$ then $f^* = \underline{\underline{a}}_X f^\circ i_Y$ is a left adjoint for f_* on sheaves. If $g : Y \to Z$ is continuous then $(gf)^* = f^*g^*$.

A map of topological spaces $f : X \to Y$ is said to be <u>relative</u> if the open sets of X are precisely the sets of the form $f^{-1}(U)$, U open in Y, i.e. the topology of X is initial for f.

<u>LEMMA</u> II.5.4. Let $f : X \to Y$ be relative, then f_* is full. If $h : P \to Q$ is a morphism in $\mathcal{P}(X,C)$ for which $f_*(h)$ is an isomorphism then h itself is an isomorphism.

It follows immediately from this that, if f is relative and $S \in \mathcal{S}(X,\underline{C})$, then there is a sheaf $S' \in \mathcal{S}(Y,\underline{C})$ such that $f^*S' = S$. In particular, if f is the inclusion map of an open set U of X in X and if S is a sheaf over X, then we call f^*S the sheaf induced by S on U and denote it by $S|U$.

Now some details on flabby sheaves and presheaves.

A presheaf $P \in \mathcal{P}(X,\underline{C})$ is said to be flabby if the restriction homomorphisms $P^X_U : P(X) \to P(U)$ are surjective for all $U \in$ Open (X). An $S \in \mathcal{S}(X,\underline{C})$ is a flabby sheaf if iS is a flabby presheaf. For example constant sheaves and sheaves of germs of continuous functions over suitable topological spaces are flabby. The property of being flabby is obviously a local property.

PROPOSITION II.5.5. Let $f : X \to Y$ be a continuous map and let P be a flabby (pre) sheaf over X, then f_*P is a flabby (pre) sheaf over Y.

PROOF. If $U \in$ Open (Y), we have the following commutative diagram :

$$\begin{array}{ccc} P(X) & \to & P(f^{-1}(U)) \\ \downarrow & & \downarrow \\ f_*P(Y) & \to & f_*P(U) \end{array}$$

where the top row is a surjective morphism, and the columns represent isomorphisms, hence the morphism at the bottom is surjective. ■

Generally, the canonical inclusion $i : \mathcal{S}(X,\underline{C}) \to \mathcal{P}(X,\underline{C})$ is only left exact, however :

PROPOSITION II.5.6. Let

$$\overline{0} \to S' \to S \to S'' \to \overline{0}$$

be an exact sequence in $\mathcal{S}(X,\underline{C})$ and suppose that S' is a flabby sheaf; then

$$\overline{0} \to iS' \to iS \to iS'' \to \overline{0}$$

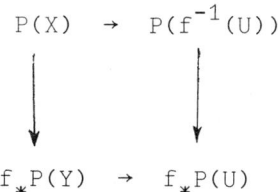

is an exact sequence of presheaves, i.e. for all $U \in \mathrm{Open}(X)$:

$$0 \to S'(U) \to S(U) \to S''(U) \to 0$$

is exact in $\underline{C}$.

PROOF. Straightforward. ■

Recall the following lemma, which is due to A. Grothendieck, cf. [13] :

LEMMA II.5.7. Let $\underline{C}$ be an abelian category, and let $\mathcal{E}$ be a class of objects of $\underline{C}$ satisfying :

1. Every object of $\underline{C}$ can be embedded in an object of $\mathcal{E}$;
2. A direct factor of an object in $\mathcal{E}$ is in $\mathcal{E}$;
3. If

$$0 \to E' \to E \to E'' \to 0$$

is an exact sequence in $\underline{C}$, with E' and E in $\mathcal{E}$, then E'' is in E; then $\mathcal{E}$ contains all injective objects of $\underline{C}$.

PROPOSITION II.5.8. An injective sheaf in $\mathcal{S}(X,\underline{C})$ is flabby.

PROOF. Apply the foregoing lemma to $\mathcal{S}(X,\underline{C})$ and take for $\mathcal{E}$ the class of flabby sheaves.

1. If $S \in \mathcal{S}(X,\underline{C})$ then S is embedded in $\tilde{S}$, where $\tilde{S}$ is the sheaf defined by $\tilde{S}(U) = \prod_{x \in U} S_x$, where $S_x = \varinjlim_{U \ni x} S(U)$ is the stalk of S at x. It is readily checked that $\tilde{S}$ is flabby.

2. Obvious, because a direct factor is also an epimorphic image.

3. Follows immediately from the following commutative diagram with

exact rows :

$$\begin{array}{ccccccccc} 0 & \to & E'(X) & \to & E(X) & \to & E''(X) & \to & 0 \\ & & {E'}^X_U \downarrow & & E^X_U \downarrow & & {E''}^X_U \downarrow & & \\ 0 & \to & E'(U) & \to & E(U) & \to & E''(U) & \to & 0 \\ & & \downarrow & & \downarrow & & & & \\ & & 0 & & 0 & & & & \end{array}$$

PROPOSITION II.5.9. Let the following sequence :

$$\overline{0} \longrightarrow M' \xrightarrow{f} M \xrightarrow{g} M'' \longrightarrow \overline{0}$$

be an exact sequence of sheaves with values in an abelian category $\underline{C}$. If M' is a flabby sheaf then the sequence is exact as a sequence of presheaves, i.e. for each $U \in \text{Open}(X)$, the sequence

$$0 \to M'(U) \xrightarrow{f(U)} M(U) \xrightarrow{g(U)} M''(U) \to 0$$

is exact in $\underline{C}$.

PROOF. It will be sufficient to give the proof in case $U = X$. Take any section $m'' \in M''(X)$ and put

$$0_{m''} = \{(U,m),\ U \in \text{Open}(X),\ m \in M(U),\ g(U)(m) = {M''}^X_U(m'')\}.$$

Let (U_0,m) be an element of $0_{m''}$ such that U_0 is maximal amongst open sets of X which appear in $0_{m''}$. If $U_0 \neq X$, let $x \in X - U_0$. The sequence :

$$0 \to M'_x \xrightarrow{f_x} M_x \xrightarrow{g_x} M''_x \to 0,$$

is exact in $\underline{C}$. Let n_x be an element of M_x which maps to ${M''}^X_x(m'')$ under g_x. There is an open neighborhood V of x and a section $n \in M(V)$ such that $g(V)(n) = {M''}^X_V(m'')$. Now :

$$g(U_0\cap V)(M^{U_0}_{U_0\cap V}(m) - M^{V}_{U_0\cap V}(n)) = M''^{U_0}_{U_0\cap V}g(U_0)(m) - M''^{V}_{U_0\cap V}g(V)(n)$$

$$= M''^{U_0}_{U_0\cap V}\, M''^{X}_{U_0}(m'') - M''^{V}_{U_0\cap V}\, M''^{X}_{V}(m'') = M''^{X}_{U_0\cap V}(m'') - M''^{X}_{U_0\cap V}(m'') = 0.$$

Therefore $M^{U_0}_{U_0\cap V}(m) - M^{V}_{U_0\cap V}(n) \in \operatorname{Im} f(U_0\cap V)$, and hence :

$$M^{U_0}_{U_0\cap V}(m) = M^{V}_{U_0\cap V}(n) + f(U_0\cap V)(m_1'),\quad m_1' \in M'(U_0\cap V).$$

Since M' is flabby, m_1' can be extended to V, $m_1' = M'^{V}_{U_0\cap V}(m_2')$ with $m_2' \in M'(V)$. Finally we obtain :

$$M^{U_0}_{U_0\cap V}(m) = M^{V}_{U_0\cap V}(n) + f(U_0\cap V)M'^{V}_{U_0\cap V}(m_2')$$

$$= M^{V}_{U_0\cap V}(n) + M^{V}_{U_0\cap V}f(V)(m_2') = M^{V}_{U_0\cap V}(n + f(V)(m_2')) = M^{V}_{U_0\cap V}(m_1)$$

where $m_1 = n + f(V)(m_2') \in M(V)$. Since M is a sheaf there exists a section $m_0 \in M(U_0\cup V)$ such that $M^{U_0\cap V}_{V}(m_0) = m_1$ and $M^{U_0\cup V}_{U_0}(m_0) = m$ but $x \in U_0\cup V$ and $U_0\cup V \in O_{m''}$ contradicts the choice of U_0. Hence $U_0 = X$. ■

PROPOSITION II.5.10. If

$$\overline{0} \to M' \to M \to M'' \to \overline{0}$$

is an exact sequence of sheaves with values in an abelian category $\underline{C}$ such that M' and M'' are flabby sheaves, then M is flabby.

PROOF. The foregoing proposition gives rise to the following exact commutative diagram in $\underline{C}$, for each $U \in \operatorname{Open}(X)$,

$$\begin{array}{ccccccccc}
0 & \to & M'(X) & \to & M(X) & \to & M''(X) & \to & 0\\
 & & \downarrow & & \downarrow & & \downarrow & & \\
0 & \to & M'(U) & \to & M(U) & \to & M''(U) & \to & 0\\
 & & \downarrow & & \downarrow & & \downarrow & & \\
 & & 0 & \to & \operatorname{Coker} M^X_U & \to & 0 & &
\end{array}$$

Hence $\operatorname{Coker} M^X_U = 0$ and M is flabby. ■

II.6. Idempotent filters

The only difference between the abstract category-theoretical set up of localization introduced in the foregoing paragraphs and localization in R-mod resides in the use of idempotent filters. The reason why things work so well in R-mod is twofold : first, it is obvious that R is a generator for R-mod, and most proofs in the module case make (implicit) use of this; next, a submodule of a module is determined by the set of its points, which allows us to introduce filters. The purpose of this paragraph is to show how an analogous situation can be created for an arbitrary Grothendieck category.

Recall that if $\underline{C}$ is a Grothendieck category with generator G, then for each $M \in \mathrm{Ob}\,\underline{C}$ there is an epimorphism $G^{(I)} \twoheadrightarrow M$ for some set I. Moreover, the canonical morphisms

$$G^{(\mathrm{Hom}_{\underline{C}}(G,M))} \to M$$

are all epic if and only if G is a generator for $\underline{C}$.
Let C be an object of $\underline{C}$ and let $q(C) = \mathrm{Hom}_{\underline{C}}(C,C)$. It is clear that q(C) is a ring, hence we get a functor

$$q = \mathrm{Hom}_{\underline{C}}(C,-) : \underline{C} \to \mathrm{Mod}-q(C).$$

One easily proves that q has a left adjoint T, and the Gabriel-Popescu theorem states that q is fully faithful iff G is a generator of $\underline{C}$, and this is equivalent to T being exact and inducing an equivalence between $\underline{C}$ and Mod-q(G)/Ker T, the quotient category of mod-q(G) relative to the localizing subcategory Ker T, which consists of all $M \in$ Mod-q(G) such that TM = 0. ∎

Let us fix a generator G for $\underline{C}$ and assume that K is a kernel functor in $\underline{C}$. We will denote by $L(G,K)$ the class of all subobjects of I in G in $\underline{C}$ with the property that G/I is K-torsion.

<u>PROPOSITION</u> II.6.1. A necessary and sufficient condition for E to be K-injective is that each diagram

$$\begin{array}{ccccc} 0 & \to & I & \to & G \\ & & \downarrow & & \\ & & E & & \end{array}$$

with $I \in L(G,K)$ can be completed commutatively.

<u>PROOF</u>. Necessity being clear, let us prove sufficiency : consider the following exact diagram in $\underline{C}$ with KN" = N" :

$$\begin{array}{ccccccccc} 0 & \to & N' & \to & N & \to & N'' & \to & 0 \\ & \phi\downarrow & & & & & & & \\ & & E & & & & & & \end{array}$$

We have to find an extension $\overline{\phi} : N \to E$ of ϕ. Consider the set of all couples (N^*,ϕ^*) with $N' \subseteq N^* \subseteq N$ (hence $K(N/N^*) = N/N^*$!) and $\phi^* : N^* \to E$ a map extending ϕ. Zorn's lemma yields a largest couple of this kind. Let us rename objects and call it (N',ϕ). We will prove by contradiction that $N' = N$. Assume the converse, then $N' \subsetneqq N$, hence there is map $\gamma : G \to N$ which does not factorize through N', i.e. $\operatorname{Im}\gamma \not\subset N'$. Let us exhibit the following commutative diagram where $I = \gamma^{-1}(N')$, $J = \operatorname{Ker}\gamma$:

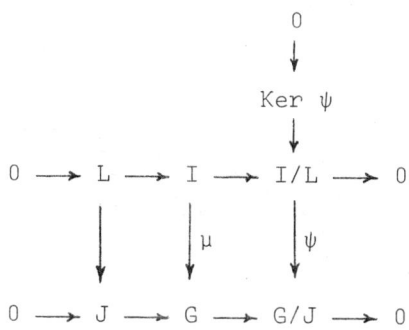

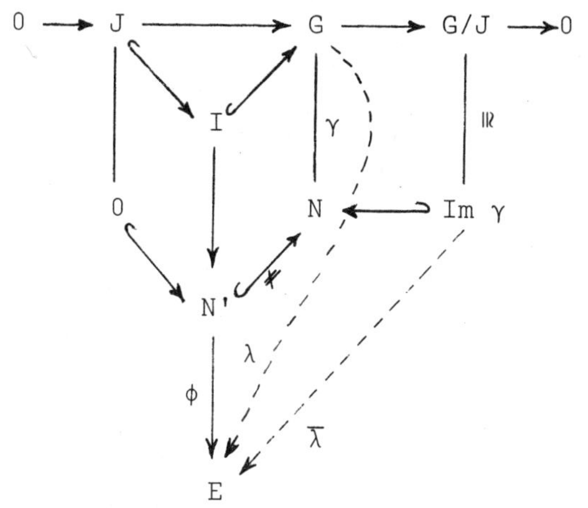

As N" is K-torsion, $I \in L(G,K)$. The map $\phi i : I \to E$ extends to a map $\lambda : G \to E$, which factorizes through G/J, as $J = \operatorname{Ker} \gamma \subseteq \operatorname{Ker} \lambda$. As $G/J \cong \operatorname{Im} \gamma$ we thus get a map $\overline{\lambda} : \operatorname{Im} \gamma \to E$. Now, as was noted before, $\operatorname{Im} \gamma \nsubseteq N'$, hence $N' \subsetneq N' + \operatorname{Im} \gamma = \check{N} \subseteq N$. On $\check{N}$ we can define $\check{\phi} : \check{N} \to E$ by $\check{\phi}|_{N'} = \phi$, $\check{\phi}|_{\operatorname{Im} \gamma} = \overline{\lambda}$. The construction of $\overline{\lambda}$ yields that $\overline{\lambda}|_{N' \cap \operatorname{Im} \gamma} = \phi|_{N' \cap \operatorname{Im} \gamma}$, hence ϕ is well defined. As $\check{\phi}$ strictly extends ϕ, this gives the expected contradiction. ■

LEMMA II.6.2. Let $I, J \in L(G,K)$ and suppose there is given $\mu \in \operatorname{Hom}_{\underline{C}}(I,G)$. Then $\mu^{-1}(J) \in L(G,K)$.

PROOF. Denote $\mu^{-1}(J)$ by L and consider the following exact diagram

$$\begin{array}{ccccccccc} & & & & & & 0 & & \\ & & & & & & \downarrow & & \\ & & & & & & \operatorname{Ker} \psi & & \\ & & & & & & \downarrow & & \\ 0 & \to & L & \to & I & \to & I/L & \to & 0 \\ & & \downarrow & & \downarrow{\scriptstyle \mu} & & \downarrow{\scriptstyle \psi} & & \\ 0 & \to & J & \to & G & \to & G/J & \to & 0 \end{array}$$

The map ψ is the quotient of μ which makes the diagram commutative. Clearly ψ is a monomorphism, so I/L can be viewed as a subobject of G/J, hence it is K-torsion. Now, in the exact sequence

$$0 \to I/L \to G/L \to G/I \to 0$$

the extremes are K-torsion, hence so is G/L, i.e. $L \in L(G/K)$. ∎

Let M be an object of $\underline{C}$, then we define

$$H_{(G,K)}(M) = \varinjlim_{I \in L(G,K)} \mathrm{Hom}_{\underline{C}}(I,M)$$

the limit being taken over the downwards directed family $L(G,K)$. Each element of $H_{(G,K)}(M)$ is represented by a map $\mu : I \to M$ for some $I \in L(G,K)$, two maps $\mu_1 : I_1 \to M$ and $\mu_2 : I_2 \to M$ being identified if there is $J \subseteq I_1 \cap I_2$ such that $J \in L(G,K)$ and $\mu_1|_J = \mu_2|_J$. We can define a pairing

$$H_{(G,K)}(M) \times H_{(G,K)}(G) \to H_{(G,K)}(M)$$

as follows. Take $g \in H_{(G,K)}(G)$ and $m \in H_{(G,K)}(M)$ and assume that they are represented by maps $\gamma : I \to G$ and $\mu : J \to M$ resp. Then we define m.g. to be the element of $H_{(G,K)}(M)$ represented by the map

$$\gamma^{-1}(J) \xrightarrow{\gamma} J \xrightarrow{\mu} M.$$

The previous lemma yields $\gamma^{-1}(J) \in L(G,K)$ and one easily checks that the definition of m.g is independent of the choice of I,J and that the pairing defined above is biadditive.
This pairing endowes $H_{(G,K)}(G)$ with a ringstructure such that $H_{(G,K)}(M)$ is a right $H_{(G,K)}(G)$-module.
One verifies that the assignment $M \rightsquigarrow H_{(G,K)}(M)$ defines a left exact functor $H_{(G,K)} : \underline{C} \to \text{Mod-}H_{(G,K)}(G)$.

On the other hand, let us reconsider the functor

$$\mathrm{Hom}_{\underline{C}}(G,-) = q_G : \underline{C} \to \mathrm{Mod}\text{-}q(G)$$

Localization at K in $\underline{C}$ yields (by the functoriality of Q_K) for each $M \in \mathrm{Ob}\ \underline{C}$ a functorial map :

$$\mathrm{Hom}_{\underline{C}}(G,M) \to \mathrm{Hom}_{\underline{C}}(Q_K(G),Q_K(M)).$$

Composing these, we get a functor

$$S_{(G,K)} : \underline{C} \to \mathrm{Mod}\text{-}S_{(G,K)}(G)$$

where

$$S_{(G,K)} = \mathrm{Hom}_{\underline{C}}(Q_K(G),Q_K(-)).$$

<u>LEMMA</u> II.6.3. For all $M \in \mathrm{Ob}\ \underline{C}$ we have $S_{(G,K)}(M) = q_G(Q_K(M))$.

<u>PROOF</u>. First let us note that

$$q_G(Q_K(M)) = \mathrm{Hom}_{\underline{C}}(G,Q_K(M)) \cong \mathrm{Hom}_{\underline{C}}(G/KG,Q_K(M)).$$

Indeed, consider the following diagram

$$\begin{array}{ccccccccc} 0 & \longrightarrow & KG & \xrightarrow{i} & G & \xrightarrow{\pi} & G/KG & \longrightarrow & 0 \\ & & & \searrow & \downarrow \phi & \swarrow \overline{\phi} & & & \\ & & & & Q_K(M) & & & & \end{array}$$

In one direction, to each $\overline{\phi} : G/KG \to Q_K(M)$ there corresponds $\overline{\phi} \circ \pi : G \to Q_K(M)$ and in the other sense we have to check that $\phi|_{KG} = \overline{0}$ as then ϕ factorizes through G/KG, but this is obvious as KG is K-torsion and $Q_K(M)$ is K-torsion-free.

Hence we assume that G is K-torsion free, and $Q_K(G) = E_K(G)$. Now, if M is faithfully K-injective and N is K-torsion free, then by definition $\mathrm{Hom}_{\underline{C}}(E_K(N),M) \cong \mathrm{Hom}_{\underline{C}}(N,M)$ as $E_K(N)/N$ is by construction K-torsion. Applying this to $Q_K(M)$ and G immediately yields the result. ■

<u>THEOREM</u> II.6.4. <u>The functors</u> $S_{(G,K)}$ <u>and</u> $H_{(G,K)}$ <u>coincide on</u> <u>K-torsion-free objects</u>.

<u>PROOF</u>. We have to prove that $H_{(G,K)}(M) = q_G(Q_K(M))$. First we construct for each $I \in L(G,K)$ a map

$$\phi_I : \mathrm{Hom}_{\underline{C}}(I,M) \to \mathrm{Hom}_{\underline{C}}(G,Q_K(M))$$

as follows : if $\psi \in \mathrm{Hom}_{\underline{C}}(I,M)$, then $\phi_I(\psi)$ is the unique map making the following diagram commutative :

$$\begin{array}{ccc} I & \xrightarrow{\psi} & M \\ {\scriptstyle i}\downarrow & & \downarrow{\scriptstyle j_K} \\ G & \xrightarrow[\phi_I(\psi)]{} & Q_K(M) \end{array}$$

where $j_K : M \to Q_K(M)$ is the localization map.
These maps are easily noticed to be compatible with the filtration, hence there is a map

$$\varinjlim_{I \in L(G,K)} \mathrm{Hom}_{\underline{C}}(I,M) \to \mathrm{Hom}_{\underline{C}}(G,Q_K(M)).$$

Conversely, each map $\xi : G \to Q_K(M)$ is induced by some $I \to M$. Indeed take $I = \xi^{-1}(M)$, then $I \in L(G,K)$, as follows from the exactness of the following commutative diagram :

$$\begin{array}{ccccccccc} 0 & \to & I & \to & G & \to & G/I & \to & 0 \\ & & \downarrow & & \downarrow{\scriptstyle \xi} & & \downarrow & & \\ 0 & \to & M & \to & Q_K(M) & \to & Q_K(M)/M & \to & 0 \end{array}$$

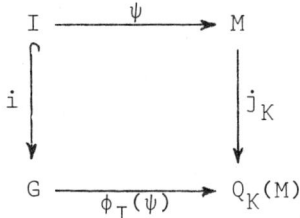

We thus get a map

$$\mathrm{Hom}_{\underline{C}}(G,Q_K(M)) \to \varinjlim_{I \in L(G,K)} \mathrm{Hom}_{\underline{C}}(I,M)$$

One now easily verifies that these maps are converses to each other, proving the assertion. ■

<u>COROLLARY</u> II.6.5. For each $M \in \mathrm{Ob}\ \underline{C}$ we have the following relations

$$\varinjlim_{I \in L(G,K)} \mathrm{Hom}_{\underline{C}}(I,M/KM) \cong q_G(Q_K(M))$$
$$\cong \mathrm{Hom}_{\underline{C}}(Q_K(G),Q_K(M)). \quad ■$$

= To indicate the significance of the "idempotent filter" $L = L(G,K)$, let us point out the following properties :

(i) If $I \in L$ and $I \subset J$, then $J \in L$;

(ii) If $I \in L$ and $J \in L$, then $I \cap J \in L$;

(iii) If $I \in L$ and $\phi \in q(G)$, then $\phi^{-1}(I) \in L$;

(iv) If $I \subset G$ and there is $K \in L$ such that for every $\phi \in q(K)$ we have $\phi^{-1}(I) \in L$, then $I \in L$.

Conversely, assume that a set L consisting of subobjects of G satisfies these conditions, then L is called an <u>idempotent filter with respect to</u> G. To such an idempotent filter corresponds a torsion theory with torsion class T consisting exactly of those objects M of $\underline{C}$ such that for every $\phi \in q(M)$ we have that $\mathrm{Ker}\,\phi \in L$. Let us prove this.

(a) T is closed under subobjects : take $i : N \hookrightarrow M$, where $M \in T$, then for every $\phi \in q(N)$, $\mathrm{Ker}\ \phi = \mathrm{Ker}(i \circ \phi) \in L$, as $i \circ \phi \in q(M)$, hence $N \in T$;

(b) T is closed under quotient objects : this follows from (i);

(c) T is closed under direct sums : easy, by (ii);

(d) T is closed under extensions : assume we have an exact sequence in $\underline{C}$:

$$0 \longrightarrow N' \xrightarrow{i} N \xrightarrow{\pi} N'' \longrightarrow 0$$

with N' and N" in T for each $\phi \in q(N)$, put $I = \mathrm{Ker}(\pi \circ \phi)$, then $I \in L$ as $\pi \circ \phi \in q(N'')$. Let $J = \mathrm{Ker}\ \phi$, then we have to show that for each $\psi \in q(I)$, we have $\psi^{-1}(J) \in L$. This can be seen as follows : consider the commutative diagram with exact top row :

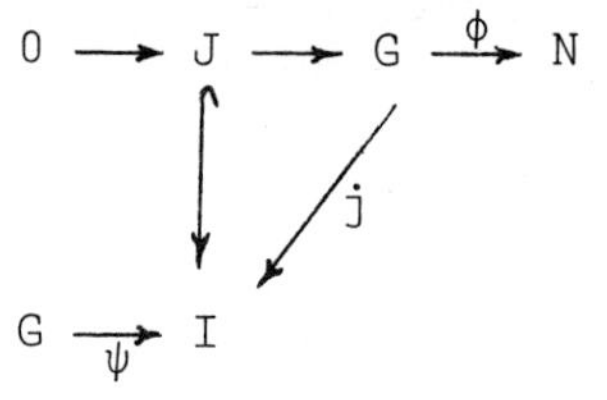

Clearly $\psi^{-1}(\mathrm{Ker}\ \phi) = \psi^{-1}(J) = \mathrm{Ker}(\phi \circ j \circ \psi)$, and as $\phi \circ j \circ \psi$ can be viewed as an element in $q(N')$, and $N' \in T$, we get $\psi^{-1}(J) \in L$, which finishes the proof.

Using II.2.3. one now gets :

<u>PROPOSITION</u> II.6.6. There is a bijective correspondence between

(a) idempotent filters with respect to any generator of $\underline{C}$;

(b) kernel functors in $\underline{C}$;

(c) hereditary torsion theories in $\underline{C}$. ∎

CHAPTER III. APPLICATION TO SHEAF THEORY

III.1. (Pre) Sheaves of Rings and Modules of Quotients: Generalities

X is a fixed topological space. Sheaves and presheaves will always be defined over X. Let R be a (pre) sheaf of rings, then, ($\pi(R)$), $\sigma(R)$ will be the category of left (pre) R-Modules, i.e. a (pre) sheaf of abelian groups M is said to be a left (pre) R-Module if for each $U \in$ Open (X), M(U) has an R(U)-module structure, compatible with restriction morphisms. A left (pre) Ideal of R is a sub (pre) sheaf of R which is a left (pre) R-Module. The prefix left is usually left out. The zero sheaf will be denoted by $\overline{0}$ in order to distinguish between $\overline{0}$ and $\overline{0}(U) = 0$. The following theorem is well known :

THEOREM II.1.1. $\pi(R)$ is a Grothendieck category and $\sigma(R)$ is a strict Giraud subcategory of $\pi(R)$.

A global section μ of R determines a subfunctor $\overline{\mu}$ of R, called Point of R, as follows : $U \in$ Open (X), $\overline{\mu}(U) = R_U^X(\mu)$. If $V \subset U$ in Open (X) then $R_V^U(\overline{\mu}(U)) = R_V^U R_U^X(\mu) = R_V^X(\mu) = \overline{\mu}(V)$. If R is a presheaf then $\overline{\mu}$ is a presheaf of sets. If $\overline{\mu}$ is a Point of R then we will write $\overline{\mu} \in R$.

Product and sum of Points are defined "sectionwise". The fact that the R_V^U are ring homomorphisms implies that product and sum of Points, defined sectionwise, yields a Point. In a similar way

Points of (pre) R-Modules, and scalar multiplication of Points may be defined. The set of Points of an $M \in \pi(R)$ "forms" a flabby subpresheaf PM of M as follows $PM(U) = \{x \in M(U), \exists y \in M(X)$ such that $M_U^X(y) = x\}$. PM is the maximal flabby presheaf contained in M. If $M \in \pi(R)$, then $PM \in \pi(PR)$, hence if R is flabby then PM is in a natural way a pre R-Module for every $M \in \pi(R)$. Obviously, if $M \in \sigma(R)$ then P i M is <u>not</u> necessarily a sheaf.

Let I be a left pre Ideal of R and $\bar{\mu} \in R$. Define $(I : \bar{\mu})(U) = \{x \in R(U), x\bar{\mu}(U) \in I(U)\}$.

<u>LEMMA</u> III.1.2. Let I be a left (pre)-Ideal of R, $\bar{\mu} \in R$. If R is a presheaf, then $(I : \bar{\mu})$ is a left pre-Ideal, if R is a sheaf, then $(I : \bar{\mu})$ is a left Ideal.

<u>PROOF</u>. Let $x \in (I : \bar{\mu})(U)$, $U \in$ Open (X). We have : $R_V^U(x\bar{\mu}(U)) \in R_V^U(I(U)) = I_V^U(I(U)) \subset I(V)$. On the other hand $R_V^U(x\bar{\mu}(U)) = R_V^U R_U^X(\mu) = R_V^U(x)R_V^X(\mu) = R_V^U(x)\bar{\mu}(V)$, therefore R_V^U restricts to a ring homomorphism $(I : \bar{\mu})(U) \to (I : \bar{\mu})(V)$, which makes $(I : \bar{\mu})$ into a (pre) R-Module because it is clear that the $(I : \bar{\mu})_V^U$ satisfy the transitivity relations and that the identity of R(U) restricts to the identity of $(I : \bar{\mu})(U)$. Assume that R and I are sheaves. Take $U \in$ Open (X) and $\{U_i\}_{i \in I} \in Cov_X(U)$. Denote by ρ_i, resp. ρ_{ij}^i, the restriction homomorphisms $R_{U_i}^U$, resp. $R_{U_i \cap U_j}^{U_i}$, of R. Since I and $(I : \bar{\mu})$ are subpresheaves of R it will cause no ambiguity if the restriction homomorphisms for these will be denoted by the same symbol. Suppose that, for each $i \in J$ we have $x_i \in (I : \bar{\mu})(U_i)$ such that $\rho_{ij}^i x_i = \rho_{ij}^j x_j$ for all $j \in J$. Since R is a sheaf, there is an $x \in R(U)$ such that $\rho_i x = x_i$. Consider $x\bar{\mu}(U)$. Then, $\rho_i(x\bar{\mu}(U)) = \rho_i(x)\rho_i(\bar{\mu}(U)) = x_i\bar{\mu}(U_i) = z_i \in I(U_i)$. The choice of x_i entails that $\rho_{ij}^i z_i = \rho_{ij}^j z_j$, therefore the fact

that I is a sheaf yields that there exists a unique $z \in I(U)$ such that $\rho_i(z) = z_i$. Because the restriction homomorphisms of R and I coincide on I, the element $x\overline{\mu}(U) - z$ of $R(U)$ maps to zero under each ρ_i, therefore $z = x\overline{\mu}(U)$ and $x \in (I : \overline{\mu})(U)$. Hence $(I : \overline{\mu})$ is a sheaf. ■

III.2. Local Kernel Functors

A class of kernel functors which arises in a natural way is given by kernel functors which may be described locally by common kernel functors in module categories. Examples of these abound; in fact, all constructions of ringed spaces in algebra, which are known to the authors are actually constructions of local kernel functors, for example Spec, Gam_k (cf. [5]), Prim_k (cf. [30]), R-Sp (cf. [10], [9]).

Suppose that for each $U \in$ Open (X) we are given a kernel functor $K(U)$ in $R(U)$-mod. Let $V \subset U$ be open in X and assume that for each choice of U,V the following conditions are satisfied :

L_1. : If $M_U \in R(U)$-mod, $M_V \in R(V)$-mod, $f : M_U \to M_V$ a semilinear map with respect to $R_V^U : R(U) \to R(V)$, then the following diagram is commutative :

$$\begin{array}{ccc} K(U)(M_U) & \xrightarrow{f_{res}} & K(V)(M_V) \\ \downarrow & & \downarrow \\ M_U & \xrightarrow{f} & M_V \end{array}$$

L_2. : If M_U, $M'_U \in R(U)$-mod, M_V, $M'_V \in R(V)$-mod, are such that we have a commutative diagram :

$$\begin{array}{ccc} M_U & \xrightarrow{f} & M_V \\ {\scriptstyle g_U}\downarrow & & \downarrow{\scriptstyle g_V} \\ M'_U & \xrightarrow{f'} & M'_V \end{array}$$

where g_U, resp. g_V, is an R(U)-, resp. R(V)-, linear map and where f_1 and f_2 are semilinear with respect to R^U_V, then the following diagram is commutative :

$$\begin{array}{ccc} K(U)(M_U) & \xrightarrow{f_{res}} & K(V)(M_V) \\ {\scriptstyle K(U)(g_U)}\downarrow & & \downarrow{\scriptstyle K(V)(g_V)} \\ K(U)(M'_U) & \xrightarrow[f'_{res}]{} & K(V)(M'_V) \end{array}$$

<u>PROPOSITION</u> III.2.1. If for every $U \in$ Open (X) a kernel functor $K(U) \in R(U)$-ker is given, such that L_1 and L_2 are satisfied, then there exists a (unique) kernel functor K in $\pi(R)$ such that for every $M \in \pi(R)$ we have $K(M)(U) = K(U)(M(U))$.

<u>PROOF</u>. If we take $f = M^U_V$ in L_1 then it follows that, assigning $K(M)(U) = K(U)(M(U))$ to $U \in$ Open (X), defines a subpresheaf K(M) of M and semilinearity of f_{res} implies that in fact $K(M) \in \pi(R)$. From L_2 we deduce that K is a subfunctor of the identity. Taking sections over $U \in$ Open (X) is an exact functor from $\pi(R)$ to R(U)-mod, hence left exactness of K(U) for all $U \in$ Open (X) entails left exactness of K. Moreover, since M/K(M) is a presheaf, we get for every $U \in$ Open (X) :

$$K(M/K(M))(U) = K(U)((M/K(M))(U)) = K(U)(M(U)/K(U)(M(U))) = 0.$$

■

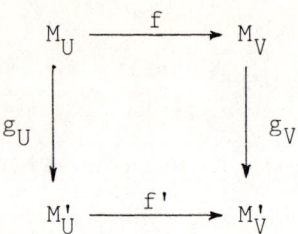

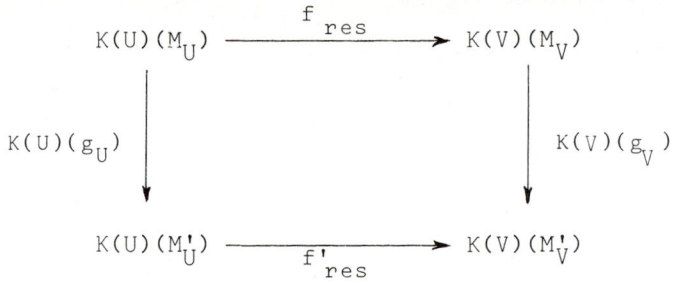

A kernel functor given by a family $\{K(U),\ U \in \text{Open }(X)\}$ where $K(U) \in R(U)\text{-Ker}$ satisfying L_1 and L_2 is said to be a <u>local</u> kernel functor in $\pi(R)$. If K is a local S-compatible kernel functor in $\pi(R)$ then K^S will be called a <u>local kernel functor in $\sigma(R)$</u>.

<u>PROPOSITION</u> III.2.2. Let $\bar{R}_V^U$ be the lattice morphism $R(U)\text{-ker} \to R(V)\text{-ker}$ induced by the ring homomorphism R_V^U. A kernel functor $K \in R\text{-ker}$ is local if and only if $\bar{R}_V^U(K(U)) \leqslant K(V)$ for all $V \subset U$ in Open (X).

<u>PROOF</u>. If K satisfies L_1 then taking $M = M'$, $f = 1_M$, yields $\bar{R}_V^U(K(U)) \leqslant K(V)$. The converse follows easily from well known facts about kernel functors in module categories. ■

A kernel functor K in $\pi(R)$ is <u>local at</u> $U \in$ Open (X) if for every $M, M' \in \pi(R)$ such that $M(U) = M'(U)$ we have that $K(M)(U) = K(M')(U)$.

<u>PROPOSITION</u> III.2.3. The following statements are equivalent :

1. K is a local kernel functor in $\pi(R)$.
2. K is local at every $U \in$ Open (X).

<u>PROOF</u>. The implication 1. ⇒ 2. is obvious. To prove the converse we have to construct kernel functors $K(U)$ in $R(U)$-ker, for every $U \in$ Open (X), such that for any $M \in \pi(R)$ we have $K(U)M(U) = K(M)(U)$. Let $M_U \in R(U)\text{-mod}$ and put $K(U)M_U = K(M)(U)$, where M is any presheaf in $\pi(R)$ such that $M(U) = M_U$. Since K is local at U there is no ambiguity in this definition. Let $(R_V^U)_* : R(V)\text{-mod} \to R(U)\text{-mod}$ be the restriction of scalars via R_V^U, $V \subset U$ in open (X). Define a presheaf $C(M_U)$ by $C(M_U)(V) = (R_U^V)_*(M_U)$ if $V \supset U$ and $C(M_U)(V) = 0$ if $V \not\supset U$; where $C(M_U)_W^V$ for $V \supset W$ is given by the zero map if $W \not\supset U$ and the relative identity

$$C(M_U)(V) \to C(M_U)(W) : rm \to R_W^V(r)m,$$

where $rm = R_U^V(r)m$ and $R_W^V(r)m = R_U^W R_W^V(r)m$, if $W \supset V$. Obviously $C(M_U)(U) = M_U$ and thus $C(M_U)$ may be used in defining $K(U)$ as before. If $M_U \to M_U'$ is a morphism in $R(U)$-mod then a morphism $C(M_U) \to C(M_U')$ in $\pi(R)$ results. Therefore it is easy to verify that $K(U)$ as defined above is a subfunctor of the identity. Let

$$0 \to M_U' \to M_U \to M_U'' \to 0$$

be an exact sequence in $R(U)$-mod then this implies exactness of the following sequences (in their corresponding categories) :

$$\overline{0} \to C(M_U') \to C(M_U) \to C(M_U'') \to \overline{0},$$

$$\overline{0} \to K(C(M_U')) \to K(C(M_U)) \to K(C(M_U'')),$$

$$0 \to K(C(M_U'))(U) \to K(C(M_U))(U) \to K(C(M_U''))(U),$$

$$0 \to K(U)(M_U') \to K(U)(M_U) \to K(U)(M_U'').$$

Hence $K(U)$ is left exact. Finally, note that

$$M_U/K(U)(M_U) = [C(M_U)/K(C(M_U))](U), \text{ so}$$

$$K(U)(M_U/K(U)(M_U)) = K(C(M_U)/K(C(M_U)))(U) = 0. \blacksquare$$

Let us also mention the following : if $\{U_i\}_{i \in I} \in \mathrm{Cov}_X(X)$, let R_i be the restriction of R to U_i and consider K_i in $\pi(R_i)$ such that for all $M \in \pi(R)$ we get :

$$K_i(M|U_i \cap U_j) \xrightarrow[\phi_{ij}]{\sim} K_j(M|U_i \cap U_j),$$

such that for each triple (i,j,k), $\phi_{ik} = \phi_{jk}\phi_{ij}$, then there exists a kernel functor K in R-ker such that $K(M)|U_i = K_i(M|U_i)$.

The proof of this fact is straightforward.

III.3. Idempotent Filters

Although the importance of idempotent filters is strongly linked to the existence of a well behaved set of underlying "points", they do become significant when related to Points of presheaves.

A set L consisting of left pre-Ideals of R is said to be an idempotent filter if :

1. If $I \in L$, $\bar{\mu} \in R$, then $(I : \bar{\mu}) \in L$.
2. If I is a left (pre) Ideal of R such that there exists a $J \in L$ such that for all $\bar{\mu} \in J$, $(I : \bar{\mu}) \in L$ then $I \in L$.

PROPOSITION III.3.1. Let L be an idempotent filter, then :

1. $I, J \in L$ yields $I \cap J \in L$.
2. $I \in L$, $I \subset J$ then $J \in L$.

PROOF. 1. Let $\bar{\mu} \in J$, then $((I \cap J) : \bar{\mu})(U) = \{x \in R(U), x\bar{\mu}(U) \in (I \cap J)(U)\} = \{x \in R(U), x\bar{\mu}(U) \in I(U) \cap J(U)\}$

$$= (I : \bar{\mu})(U) \cap (J : \bar{\mu})(U) = (I : \bar{\mu})(U).$$

Therefore $((I \cap J) : \bar{\mu}) = (I : \bar{\mu}) \in L$ for every $\bar{\mu} \in J$, hence $I \cap J \in L$.

2. Take $\bar{\mu} \in I$; then for every $U \in$ Open (X) and every $x \in R(U)$, $x\bar{\mu}(U) \in I(U) \subset J(U)$. Hence $(J : \bar{\mu})(U) = R(U)$ for all $U \in$ Open (X), hence $(J : \bar{\mu}) = R$. Taking $\bar{\mu} = 0$ yields $R \in L$, the foregoing then implies $J \in L$. ■

Let K be a kernel functor in $\pi(R)$. Define $L(K) = \{I$ left pre-Ideal of R such that $K(R/I) = R/I\}$.

<u>LEMMA</u> III.3.2. Let $M \in \pi(R)$, $\overline{\mu}$ a Point of M, then the following statements are equivalent :

1. $\overline{\mu} \in K(M)$
2. There is an $I \in L(K)$ such that $I\overline{\mu} = \overline{0}$, i.e. $I(U)\overline{\mu}(U) = 0$ for every $U \in$ Open (X).

<u>PROOF</u>.

1. ⇒ 2. Put $I = \text{Ann}\,\overline{\mu} = (\overline{0} : \overline{\mu})$. Then $(R/I)\,(U) = R(U)/I(U)$, i.e. $R/I \cong R\overline{\mu}$.

Since $R\overline{\mu} \subset K(M)$, this yields $K(R/I) = R/I$ and thus $I \in L(K)$.

2. ⇒ 1. If $I\overline{\mu} = \overline{0}$, then $I \subset (\overline{0} : \overline{\mu})$ and thus $R\overline{\mu}$ is isomorphic to an homomorphic image of R/I. Therefore $R_{\overline{\mu}}$ is K-torsion, whence $R\overline{\mu} \subset K(M)$. ■

The following properties of $L(K)$ are easily verified :

1°. For every $\overline{\mu} \in R$, $I \in L(K)$ we have $(I : \overline{\mu}) \in L(K)$.

2°. $I \in L(K)$ and $I \subset J$ implies $J \in L(K)$.

3°. If $I,J \in L(K)$, then $I \cap J \in L(K)$.

4°. If $I \in L(K)$ and $J \subset I$ such that $K(I/J) = I/J$ then $J \in L(K)$.

However $L(K)$ does not necessarily satisfy condition 2 for an idempotent filter, unless additional flabbiness conditions are imposed, see Lemma III.6.1.

If R is a sheaf of rings, K a $\sigma(R)$-compatible kernel functor in $\pi(R)$, then we define $L(K^s) = \{I \text{ left Ideal of } R \text{ such that } K(R/I) = R/I\}$.

<u>PROPOSITION</u> III.3.3. Let K be a $\sigma(R)$-compatible kernel functor in $\pi(R)$ inducing K^s in $\sigma(R)$, then : $L(K^s) = \underline{\underline{a}}L(K)$.

PROOF. This follows directly from the definition of K^s and Lemma II.4.6.,2., (because any left pre-Ideal of R is a separable presheaf as R is). ■

III.4. Ring Objects of Quotients

We include a little disgression here, and introduce reductions of rings and kernel functors. These topics have been studied in more detail in [21], [31]. Consider a ring homomorphism $\pi : R_1 \to R_2$, and suppose that π is onto. In that case, the kernel functor in R_2-mod $\bar{\pi}\kappa_1$ associated to $\kappa_1 \in R_1$-ker is given by $L(\bar{\pi}\kappa_1) = \pi L(\kappa_1)$. If R_1, R_2 are equipped with the topologies derived from κ, $\bar{\pi}\kappa$ resp. then $\pi : R_1 \to R_2$ is a continuous morphism of topological rings and it is also final, i.e. it maps open sets to open sets. If $\mathrm{Ker}\ \pi \subset \kappa_1(R_1)$ then π is said to be a torsion reduction with respect to κ_1, short : κ_1-reduction. If $\kappa_2 \in R_2$-ker is such that $\kappa_2 \geq \bar{\pi}\kappa_1$, then π is open with respect to the topologies associated to κ_1, κ_2.

Let $M_1 \in R_1$-mod, $M_2 \in R_2$-mod and let $\pi : R_1 \to R_2$ be a κ_1-reduction. An R_1-epimorphism $g : M_1 \to M_2$ is said to be a κ_1-reduction of M_1 if $\bar{\pi}\kappa_1(M_2) = g(\kappa_1(M_1))$. If moreover $\mathrm{Ker}\ g \subset \kappa_1(M_1)$ then g is said to be a torsion reduction of M_1. From [21] we recall :

LEMMA III.4.1. Let $g : M_1 \to M_2$ be a torsion reduction of M_1 then, if M_1 is κ_1-projective, then M_2 is $\bar{\pi}\kappa_1$-projective. Furthermore, let κ_1 be a t-functor, then if M_1 is faithfully κ_1-injective, M_2 is faithfully $\bar{\pi}\kappa_1$-injective too.

COROLLARY. If $\pi : R_1 \to R_2$ is a κ_1-reduction, where $\kappa_1 \in R_1$-ker is a t-functor, then $\pi_* Q_{\bar{\pi}\kappa_1} = Q_{\kappa_1}\pi_*$.

It follows that if $\pi : R_1 \to R_2$ is an epimorphism and if $\kappa_2 \geq \overline{\pi}\kappa_1$, then we have an R_2-linear morphism $Q_{\overline{\pi}\kappa_1}(R_2) \to Q_{\kappa_2}(R_2)$ where $\pi_* Q_{\overline{\pi}\kappa_1}(R_2) = Q_{\kappa_1}(R_1)$; this may be considered as giving a semilinear morphism $\tilde{\pi} : Q_{\kappa_1}(R_1) \to Q_{\kappa_2}(R_2)$, with respect to $\pi : R_1 \to R_2$. Note that $\tilde{\pi}$ is a ring morphism, indeed this follows from the fact that $Q_{\kappa_2}(R_2)$ is faithfully $\overline{\pi}\kappa_1$-injective, cf. corollary 1 to Proposition I.1.10.

Throughout the sequel of this section we assume that R is a flabby presheaf of rings and $K \in R$-ker is a local kernel functor reducing R, i.e. $\operatorname{Ker} R_V^U \subset K(U)(R(U))$ for every $U,V \in \operatorname{Open}(X)$, $U \supset V$. All results are derived in $\pi(R)$, however similar results hold in $\sigma(R)$ and to prove them it is sufficient to use the corollary to Proposition II.4.10.

If K reduces R then $K(U)(R(U))$ is exactly the inverse image of $K(U)(R(V))$ under R_V^U. Since K is local we have that $\overline{R}_V^U\, K(U) \leq K(V)$ and thus the morphism $(R/K(R))_V^U$ which decomposes as :

$$R(U)/K(U)(R(U)) \stackrel{\cong}{\to} R(V)/K(U)(R(V)) \to R(V)/K(V)(R(V)),$$

is surjective. This entails that $R/K(R)$ is a flabby presheaf of rings and the canonical epimorphism $\overline{j}_K : R \to R/K(R)$ determines a kernel functor $\overline{j}_K K$ in $\pi(R/K(R))$ as follows $(\overline{j}_K K)(U) = \overline{j}_{K(U)} K(U)$, where $j_{K(U)}$ is the canonical morphism $R(U) \to R(U)/K(U)(R(U))$. Obviously, $\overline{j}_K K$ is local. The fact that $K(V) \geq \overline{R}_V^U K(U)$ implies that we have canonical ring homomorphisms :

$$Q_{K(U)}(R(U)) \to Q_{K(V)}(R(V))$$

These extend $(R/K(R))_V^U$ and thus $Q_{K(U)}(R(U)) \cong Q_{K(U)}(R(V))$ for all $U \supset V$ in Open (X). In this way we obtain a Ring $Q \in \pi(R)$, which is clearly K-torsion free and also K-injective, as is easily seen.

The construction also implies that $Q/(R/K(R))$ is K-torsion. Uniqueness of $Q_K(R)$, as constructed before, yields $Q = Q_K(R)$.

THEOREM III.4.2. Let K be a local kernel functor that reduces R, then $Q_K(R)$ is a Ring. The Ring structure of $Q_K(R)$ is uniquely determined by its R-Module structure.

Note : $Q_K(R)$ is in a natural way an $R/K(R)$-Module containing $R/K(R)$. Moreover $\overline{j}_K K(Q_K(R)) = \overline{0}$ and $Q_K(R)$ is $\overline{j}_K K$-injective by construction, hence : $Q_K(R) = Q_K(R/K(R)) \cong Q_{\overline{j}_K K}(j_K(R))$. We may therefore say that j_K is a torsion reduction, generalizing the analoguous concept in the ring-theoretic case. Note too, that if K reduces R and $K(X)(R(X)) = 0$, then R has to be the constant presheaf. So if R is a (pre) sheaf of prime rings then the $Q_K(R)_V^U$ are monomorphisms, cf. Proposition I.3.2. for a well known example.

THEOREM III.4.3. Let K be a local kernel functor reducing R and let $M \in \pi(R)$. Then $Q_K(M)$ is given by $Q_K(M)(U) = Q_{K(U)}(M(U))$, $U \in \text{Open}(X)$. $Q_K(M)$ is in a natural way a $Q_K(R)$-Module and this structure is uniquely determined by its R-Module structure.

PROOF. Put $Q(U) = Q_{K(U)}(M(U))$. Denote $\overline{R}_V^U K(U)$ by $K'(U)$. Let $V \subset U$ in Open (X) and consider $Q_{K(U)}(M(V))$. First note that the $R(U)$-module structure of $Q_{K(U)}(M(V))$ is in fact given as an $R(V)$-module structure via $R(U) \to R(V)$; indeed if $\lambda \in \text{Ker } R_V^U$ and $x \in Q_{K(U)}(M(V))$ then $\lambda x = 0$ because $\text{Ker } R_V^U \subset K(U)(R(U))$, hence

$$\lambda x \in K(U)(Q_{K(U)}(M(V)) = 0.$$

Now $Q_{K(U)}(M(V))$ is an $R(V)$-module which is $K(U)$-injective, therefore it is also $K'(U)$-injective; indeed, if we are given a diagram in R(V)-mod :

$$\begin{array}{ccl} 0 & & \\ \downarrow & & \\ N' & \xrightarrow{f} & Q_{K(U)}(M(V)) \\ \downarrow & \nearrow_{g} & \\ N & & \\ \downarrow & & \\ N/N' & & \\ \downarrow & & \\ 0 & & \end{array}$$

where N/N' is $K'(U)$-torsion, then N/N' is $K(U)$-torsion as an $R(U)$-module, therefore f, which is also $R(U)$-linear extends to an $R(U)$-linear $g : N \to Q_{K(U)}(M(V))$, but since N as well as $Q_{K(U)}(M(V))$ are $R(V)$-modules we have that g is $R(V)$-linear.

Note that $M(V)/K(U)(M(V)) = M(V)/K'(U)(M(V))$.

Since $Q_{K(U)}(M(V))/j_{K(U)}(M(V))$ is $K(U)$-torsion it is $K'(U)$-torsion as an $R(V)$-module. Moreover, $Q_{K(U)}(M(V))$ is clearly $K'(U)$-torsion free, so resuming all this : $Q_{K(U)}(M(V)) = Q_{K'(U)}(M(V))$. It follows then that $M_V^U : M(U) \to M(V)$, extends to a unique $R(U)$-linear morphism :

$$Q_V^U : Q_{K(U)}(M(U)) \to Q_{K(U)}(M(V)) = Q_{K'(U)}(M(V)) \to Q_{K(V)}(M(V)).$$

One easily checks that this defines a presheaf Q containing $M/K(M) = M'$, in such a way that Q/M' is K-torsion while Q is K-torsion free. To check whether Q is K-injective it suffices to take sections over $U \in$ Open (X), hence by the uniqueness of $Q_K(M)$ with the listed properties it follows that $Q \cong Q_K(M)$. Local considerations easily yield the uniqueness of the $Q_K(R)$-Module structure as desired. ■

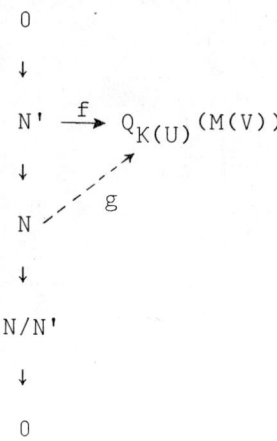

If K is a local $\sigma(R)$-compatible kernel functor in $\pi(R)$, where R is a sheaf of rings, then we say that K^s reduces R if K reduces iR. From the corollary to Proposition II.4.10. it follows immediately that we have :

PROPOSITION III.4.4. Let R be a flabby sheaf of rings and let K be a local $\sigma(R)$-compatible kernel functor in $\pi(R)$ reducing R. Then $Q_{K^s}(R)$ is a sheaf of rings and this Ring-structure is uniquely determined by its R-Module structure. For any $M \in \sigma(R)$, $Q_{K^s}(M)$ is in a natural way a $Q_K(R)$-Module; moreover, for all $U \in$ Open (X) we have, $Q_K(M)(U) = Q_{K(U)}(M(U))$.

III.5. T-Functors in R-Ker

THEOREM III.5.1. *Let R be a flabby presheaf, K a local kernel functor in $\pi(R)$ reducing R then the following statements are equivalent* :

1. *Every $M \in \pi(Q_K(R))$ is faithfully K-injective*
2. *The functor Q_K is exact and commutes with direct sums.*
3. *Every $M \in \pi(Q_K(R))$ is K-torsion free.*

PROOF. 1. ⇒ 2. Consider the exact sequence in $\pi(R)$:

$$(*) \qquad \overline{0} \to N \to M \to M/N \to \overline{0}$$

By left exactness of Q_K, $Q_K(M)/Q_K(N)$ is embedded into $Q_K(M/N)$. If one checks that $\mathrm{Ker}(M/N \to Q_K(M)/Q_K(N))$ is K-torsion, then it follows that $Q_K(M/N) = Q_K(M)/Q_K(N)$ because, by 1., $Q_K(M)/Q_K(N)$ is faithfully K-injective.

Now, $\quad \overline{0} \to K(N) \to K(M) \to K(M/N) \quad$ is exact.

Moreover,

$$(M/K(M))/(N/K(N)) \cong M/N + K(M) \cong (M/N)/(K(M)/K(N)),$$

so the kernel of

$$(M/K(M))/(N/K(N)) \to (M/N)/K(M/N)$$

is isomorphic to $K(M/N)/(K(M)/K(N))$, hence K-torsion. This entails that

$$Q_K(M/N) = Q_K((M/K(M))/(N/K(N))),$$

yielding that in (*) we may consider M to be K-torsion free. Then we have an exact sequence in $\pi(R)$:

$$\overline{0} \to M \cap Q_K(N) \to M \to Q_K(M)/Q_K(N)$$

and thus also the following sequence is exact :

$$\overline{0} \to M \cap Q_K(N)/N \to M/N \to Q_K(M)/Q_K(N).$$

Since Q_K is left exact and as $\overline{0} = Q_K(M \cap Q_K(N)/N)$ we find that $Q_K(M/N)$ injects into $Q_K(M)/Q_K(N)$, the latter being faithfully K-injective. This proves that Q_K is right exact. Since $\bigoplus_i Q_K(M_i)$ is faithfully K-injective, while $\bigoplus_i Q_K(M_i)/\bigoplus M_i$ is K-torsion, we get $Q_K(\bigoplus_i M_i) = \bigoplus_i Q_K(M_i)$.

2. ⇒ 3. Let

$$0 \to M_U' \to M_U \to M_U'' \to 0$$

be an exact sequence in R(U)-mod. Using the construction of $C(M_U)$ as in Proposition III.2.3. we obtain an exact sequence in $\pi(R)$

$$\overline{0} \to C(M_U') \to C(M_U) \to C(M_U'') \to \overline{0}$$

Then applying Q_K we get :

$$\overline{0} \to Q_K(C(M_U')) \to Q_K(C(M_U)) \to Q_K(C(M_U'')) \to \overline{0}$$

and taking sections under U again this yields

$$0 \to Q_{K(U)}(M'_U) \to Q_{K(U)}(M_U) \to Q_{K(U)}(M''_U) \to 0$$

using Theorem IV.4.3.). Hence $Q_{K(U)}$ is exact for every $U \in \mathrm{Open}(X)$. Since C obviously commutes with direct sums, one derives in a similar way that $Q_{K(U)}$ commutes with direct sums in R(U)-mod. Thus K(U) is a t-functor. So if $M \in \pi(Q_K(R))$ then M(U) is a $Q_{K(U)}(R(U))$-module where $K(U) \in R(U)$-ker is a t-functor, i.e. $K(U)(M(U)) = 0$ what implies $K(M) = 0$, thus 3.

3. ⇒ 1. If $M \in \pi(Q_K(R))$ is K-torsion free then $j_K : M \to Q_K(M)$ is an injective morphism. Consider :

$$\overline{0} \to M \to Q_K(M) \to Q_K(M)/j_K(M) \to \overline{0}.$$

Since $Q_K(M)/j_K(M)$ is in $\pi(Q_K(R))$ it is K-torsion free but then $Q_K(M) = M$ and 1. follows. ■

PROPOSITION IV.5.2. Let R be a flabby presheaf, K a kernel functor satisfying the conditions of Theorem IV.5.1. then :

1) For each $I \in L(K)$, $Q_K(R)j_K(I) = Q_K(R)$, j_K being the canonical $j_K : R \to Q_K(R)$.
2) The canonical inclusion $i_K : \pi(R,K) \to \pi(R)$, where $\pi(R,K)$ is the full subcategory of all faithfully K-injective objects of $\pi(R)$, has a right adjoint.

PROOF. 1). Since $Q_K(R)j_K(I)$ is in $\pi(Q_K(R))$ it is in $\pi(R,K)$ by the foregoing Theorem and the assumptions on K. Therefore $Q_K(R)j_K(I) = Q_K(j_K(I)) = Q_K(I)$, but obviously $I \in L(K)$ implies $Q_K(I) = Q_K(R)$. One easily constructs a right adjoint using the right adjoint of $j_K^* \circ i_K$ which is an equivalence of categories. ∎

It is now an easy exercise to prove all the results derived so far in case one considers $\sigma(R)$ instead of $\pi(R)$. In case $\sigma(R)$-compatible kernel functors are being considered this may be done immediately, using Proposition III.4.10. Any K satisfying any of the conditions of Theorem III.5.1. is said to be a T-functor in $\pi(R)$. If K is such then K(U) is a t-functor in R(U)-mod for all $U \in$ Open (X), and conversely. One easily derives the following : K is a T-functor if and only if $Q_K(M) \cong Q_K(R) \underset{R}{\otimes} M$ for every $M \in \pi(R)$. However, if K is a T-functor in $\sigma(R)$ then this does not work because the K(U) need not be t-functors. However, we have :

<u>THEOREM</u> III.5.3. <u>Let</u> K <u>be a local kernel functor in</u> $\sigma(R)$ <u>reducing</u> R, R <u>a flabby sheaf of rings</u>, <u>then the following statements are equivalent</u> :

1. K <u>is a</u> T-<u>functor in</u> $\sigma(R)$
2. $Q_K(M) \cong Q_K(R) \underset{R}{\otimes} M$ <u>for all</u> $M \in \sigma(R)$.

<u>PROOF</u>. For a generator of $\sigma(R)$ we may take $G = \bigoplus_{U \in \text{Open }(X)} R_U$ where $R_{U,x} = 0$ if $x \notin U$ and $R_{U,x} = R_x$ is $x \in U$. We have a canonical morphism :

$$Q_K(R) \underset{R}{\otimes} G \to Q_K(G).$$

Calculation of stalks yields :

$$(Q_K(R) \underset{R_x}{\otimes} G)_x = Q_K(R)_x \underset{R_x}{\otimes} G_x = \bigoplus_U Q_K(R)_x \underset{R_x}{\otimes} R_{U,x},$$

$$Q_K(G)_x = \bigoplus_U Q_K(R_U)_x.$$

1°. If $x \in U$, take any $V \in$ Open (X), $x \in V \subset U$ then $Q_K(R)(V) = Q_K(R_U)(V)$

because $R_U(V) = R(V)$, as is easily seen from

$$Q_K(R_U)(V) = Q_{K(V)}(R_U(V)) = Q_{K(V)}(R(V)).$$

Therefore $Q_K(R)_x = Q_K(R_U)_x$, $R_{U,x} = R_x$ for every $x \in U$, yields $Q_K(R)_x \otimes_{R_x} R_{U,x} \cong Q_K(R_U)_x$ for every $x \in U$.

2°. If $x \notin U$, then $R_{U,x} = 0$, so it will be sufficient to prove that $Q_K(R_U)_x = 0$. It is well known that the inductive limit of unitary rings and unitary morphisms $\{A_\alpha, \phi_{\alpha\beta}, \alpha \leqslant \beta\}$ is zero if and only if there is a γ such that $A_\gamma = 0$ and $A_\beta = 0$ for all $\beta \geqslant \gamma$. So the fact that $R_{U,x} = \varinjlim_{x \in V \subset U} R_U(V) = 0$ implies that for some $V \in \mathrm{Open}(X)$, $x \in V \subset U$, we have that $R_U(W) = 0$ for all $W \subset V$. Now,

$$Q_K(R_U)(W) = Q_{K(W)}(R_U(W)) = 0$$

for all $W \subset V$ containing x, hence

$$Q_K(R_U)_x = \varinjlim_{x \in W} Q_K(R_U)(W) = 0.$$

Combining 1° and 2° we have proved that $Q_K(R) \otimes_R G \cong Q_K(G)$. Since G is a generator for $\sigma(R)$, we have a resolution

$$G^{(J)} \xrightarrow{\ \phi\ } G^{(I)} \xrightarrow{\ \psi\ } M \to \overline{0}$$

for every $M \in \sigma(R)$. Right exactness of $Q_K(R) \otimes_R -$ yields an exact sequence

$$Q_K(R) \otimes_R G^{(J)} \to Q_K(R) \otimes_R G^{(I)} \to Q_K(R) \otimes_R M \to \overline{0}.$$

Therefore, exactness of Q_K and the fact that it commutes with direct sums (because K is a T-functor) yields a diagram :

$$\begin{array}{ccccccc}
Q_K(G)^{(J)} & \xrightarrow{Q_K(R)\,\otimes\,\phi} & Q_K(G)^{(I)} & \xrightarrow{Q_K(R)\,\otimes\,\psi} & Q_K(R)\otimes M & \rightarrow & \overline{0} \\
\Big|\cong & & \Big|\cong & & \Big\downarrow \mu & & \\
Q_K(G)^{(J)} & \xrightarrow[Q_K(\phi)]{} & Q_K(G)^{(I)} & \xrightarrow[Q_K(\psi)]{} & Q_K(M) & \longrightarrow & \overline{0}
\end{array}$$

where the left hand square is easily seen to be commutative since $Q_K(\phi)$ is $Q_K(R)$-linear. It follows that μ is an isomorphism too. The converse follows directly from the properties of the tensor product. ■

PROPOSITION III.5.4. Let $\mathcal{B}$ be a basis for the topology on X. Let K be a local $\sigma(R)$-compatible kernel functor reducing iR such that K(U) is a t-functor for every $U \in \mathcal{B}$, then K^S is a T-functor in $\sigma(R)$.

PROOF. Let $S \in \sigma(Q_K(R))$, then $K(iS)(U) = K(U)((iS)(U))$. Since $(iS)(U)$ is a $Q_K(R)(U) = Q_{K(U)}(R(U))$-module, and since K(U) is a t-functor, $K(U)((iS)(U)) = 0$ for all $U \in \mathcal{B}$. Therefore, because S is a sheaf, $K(iS) = \overline{0}$ or $K^S(S) = \overline{0}$, thus K^S is a T-functor in $\sigma(R)$. ■

A kernel functor with the property mentioned in the foregoing proposition is said to be a T($\mathcal{B}$)-functor. The following result strengthens what has been proven for general T-functors :

PROPOSITION III.5.5. Let K be a T($\mathcal{B}$)-functor, then :

1. Every left Ideal of $Q_{K^S}(R)$ is generated by a left Ideal of R.
2. $Q_{K^S}(S) = Q_{K^S}(R) \underset{R}{\otimes} S$ for every $S \in \sigma(R)$.

PROOF. 1. Let $I \subset Q_{K^S}(R)$ be a left Ideal and put $J = I \cap j_{K^S}(R)$. Then I/J is K^S-torsion, i.e. $iI/iQ_{K^S}(R)iJ$ is K-torsion. Thus $I/\underline{\underline{a}}(iQ_{K^S}(R)iJ)$ is K^S-torsion and a $Q_{K^S}(R)$-Module hence $I = \underline{\underline{a}}(iQ_{K^S}(R)iJ)$.
2. Immediate from Proposition III.5.4.

Examples of $T(\mathcal{B})$-functors are given by Spec A, where A is a commutative ring or a Zariski central ring, cf. Proposition I.5.12.

III.6. Pointwise Localization of Presheaves

Sectional representation, which is develloping into a flourishing branch of mathematics in its own right, is mainly concerned with global sections of sheaves. Our motivation for considering so-called "pointwise localization" here, is that this localization technique is such that its effect on presheaves may be studied by looking at the global sections. Hence application of this localization technique in sectional representation theory is not far away. Moreover, the construction of rings and modules of quotients works very well in the cases considered up to now and we will even be able to drop the rather restrictive condition that the kernel functor K should reduce R.

Throughout this chapter, R is a flabby Ring and all filters will be (left) idempotent. The following lemmas are to show that kernel functors $K \in$ R-ker such that $\mathcal{L}(K)$ is an idempotent filter, abandon.

LEMMA III.6.1. The filter $\mathcal{L}$ is idempotent if and only if it has a filterbasis of flabby left pre-Ideals.

PROOF. Suppose that $\mathcal{L}$ is idempotent and pick $I \in \mathcal{L}$. For every

$\overline{\mu} \in I$, we have $\overline{\mu} \in PI$ and $(PI : \overline{\mu}) = R \in L$ therefore $PI \in L$. Conversely, let L have a filterbasis of flabby left pre-Ideals and suppose that I is a left pre-Ideal of R such that $(I : \overline{\mu}) \in L$ for all $\overline{\mu} \in J \in L$. Then $I + PJ/I$ is flabby and every Point of $I + PJ/I$ may be annihilated by an element of L. We obtain the following exact sequence in $\pi(R)$ of flabby presheaves :

$$\overline{0} \to I + PJ/I \to R/I \to R/I + PJ \to \overline{0}$$

where both extremes are flabby and have the property that each Point may be annihilated by an element of L. Let $\overline{e}$ be the image of $\overline{1} \in R$ in R/I, then $I_1\overline{e} \subset I + PJ/I$ forsome $I_1 \in L$, hence $(\text{Ann}\ \overline{e} : \overline{\mu}) \in L$ for every $\overline{\mu} \in I_1 \in L$ or $\text{Ann}\ \overline{e} \in L$. Finally, $(\text{Ann}\ \overline{e}).\overline{1} \subset I$ and this yields $I \in L$. ∎

LEMMA III.6.2. If $K \in R\text{-Ker}$ is local then $L(K)$ is an idempotent filter.

PROOF. Let $I \in L(K)$, by definition R/I is then K-torsion. Since R/I is flabby, the restriction maps $(R/I)^X_U$, $U \in \text{Open}(X)$ are surjective, so we get surjective $\pi(R)$-morphisms f and g,

$$(R/I)(X) \xrightarrow{f} R(U)/R^X_U(I(X)) \xrightarrow{g} (R/I)(U).$$

Since K is local, $\overline{R}^X_U\ K(X) \leq K(U)$, $U \in \text{Open}(X)$. Furthermore, since R/I is K-torsion we have :

$$K(X)(R(X)/I(X)) = R(X)/I(X).$$

Surjectivity of f yields that $R(U)/R^X_U(I(X))$ is $K(X)$-torsion hence $K(U)$-torsion as an $R(U)$-module. Clearly the presheaf R/PI is determined by

$$(R/PI)(U) = R(U)/R^X_U(I(X)),$$

hence it is K-torsion whence $PI \in \mathcal{L}(K)$ follows. The foregoing lemma completes the proof. ■

A presheaf of R-modules $M \in \pi(R)$ is said to be <u>PK-free</u> if $PK(M) = \overline{0}$, M is said to be <u>PK-torsion</u> if $PK(M) = M$, i.e. if M is flabby and F-torsion. Note that PK is a functor but it is <u>not</u> left exact, hence PK is <u>not</u> in R-Ker. Still, we define <u>PK-injectivity</u> and <u>faithful PK-injectivity</u> as in Section II.2., upon replacing K by PK. As a functor PK is left semi-exact while P is right semi-exact as well. Although not left-exact PK satisfies a certain idempotency condition (actually a "radical" property),

$$PK(M/PK(M)) = \overline{0} \text{ for every } M \in \pi(R).$$

<u>PROPOSITION</u> III.6.3. Let $E \in \pi(R)$, then equivalently :

1. E is PK-injective and PK-free
2. E is faithfully PK-injective.

<u>PROOF</u>. Consider the following diagram in $\pi(R)$:

$$\begin{array}{ccccccccc} \overline{0} & \to & N & \to & M & \xrightarrow{\pi} & M/N & \to & \overline{0} \\ & & {\scriptstyle f}\downarrow & \swarrow_{g} & & & & & \\ & & E & & & & & & \end{array}$$

where the top row is exact and M/N is PK-torsion.
Assume 1. If $g \neq g'$ both extend f to M, then for some $U \in \text{Open }(X)$ and some $x \in M(U)$, $g(U)(x) \neq g'(U)(x)$. Let $y \in (M/N)(U)$ be the image of x under $\pi(U)$. Since M/N is supposed to be flabby, there is a Point $\overline{U} \in M/N$ such that $\overline{\nu}(U) = y$. Since π is onto, we may take $\overline{\mu} \in M$ such that $\overline{\nu}(X) = \pi(X)\overline{\mu}(X)$. As π respects presheaf restriction morphisms we obtain $\overline{\mu}(U) = x + z$ for some $z \in N(U)$, thus :

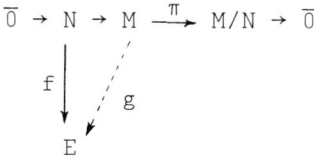

$$g(U)\overline{\mu}(U) = g(U)(x) + g(U)(z) = g(U)(x) + f(U)(z)$$

$$g'(U)\overline{\mu}(U) = g'(U)(x) + g'(U)(z) = g'(U)(x) + f(U)(z).$$

Therefore, the Points $g\overline{\mu}$ and $g'\overline{\mu}$ of E do not coincide. Now $I\overline{\nu} = \overline{0}$ for some $I \in L(K)$, i.e. $I\overline{\mu} \subset N$. This entails that $g\overline{\mu} - g'\overline{\mu}$ is a non-zero Point of E which is annihilated by $I \in L(K)$; by Lemma III.3.2., $g\overline{\mu} - g'\overline{\mu} \in K(E) = 0$, contradiction.
Assume 2. Since PK(E) is PK-torsion, the zero map $\overline{0} \to E$ extends in a unique way to a morphism $PK(E) \to E$; hence the latter has to be the zero map, hence $PK(E) = \overline{0}$. ■

PROPOSITION III.6.4. Let E be a presheaf of R-modules, E is PK-injective if and only if every R-morphism $h : I \to R$ with $I \in L(K)$, extends to $g : R \to E$, such that the following diagram is commutative :

$$\begin{array}{ccccc} \overline{0} & \to & I & \to & R \\ & & {\scriptstyle h}\downarrow & \swarrow {\scriptstyle g} & \\ & & E & & \end{array}$$

PROOF. Necessity of the given condition is obvious. For sufficiency, consider the following exact sequence in $\pi(R)$:

$$\overline{0} \to N \to M \xrightarrow{\pi} M/N \to \overline{0}$$

where M/N is PK-torsion. Let $f : N \to E$ be a given morphism. Choose a maximal element (N',f') in the set

$$\{(N'',f''),\ N \subset N'' \subset M,\ f'' : N'' \to E \text{ with } f''|N = f\}.$$

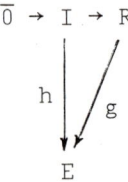

Since M/N is flabby, N + PM maps onto M/N under π, hence $M = N + PM$. So, if we succeed in proving that N' contains all Points of M then N' ⊃ N implies N' = M. We may, without loss of generality, assume that N = N'. Let $\overline{\mu} \in M$, put $I = (\overline{0} : \pi\overline{\mu})$. Then $R/I \cong R\pi\overline{\mu}$. Clearly $I\overline{\mu} \subset N$ and $I \subset (N : \overline{\mu})$. Define $h(V) : I(V) \to E(V)$ by $a \to f(V)(a\overline{\mu}(V))$. This yields an R-linear presheaf morphism h and by hypothesis there exists a morphism g^* making the following diagram commute :

$$\begin{array}{ccccc} \overline{0} & \to & I & \to & R \\ & & h\big\downarrow & \swarrow g^* & \\ & & E & & \end{array}$$

The unit Point $\overline{1}$ of R maps to $\overline{\eta} \in E$ under g^*. For U ∈ Open (X), $a \in I(U)$, we have :

$$h(U)(a) = g^*(U)(a) = a\, g^*(U)(1) = a\overline{\eta}(U).$$

Let $N + R\overline{\mu}$ be the R-Module generated by N and $\overline{\mu}$ in M. Define $g : N + R\overline{\mu} \to E$ by putting :

$$g(U)(y + a\overline{\mu}(U)) = f(U)(y) + a\overline{\eta}(U),$$

for $y \in N(U)$, $a(U) \in R(U)$. This is well defined because if $a \in I(U)$, i.e. $a\overline{\mu}(U) \in N(U)$, then :

$$f(U)(a\overline{\mu}(U)) = h(U)(a) = g^*(U)(a) = a\overline{\eta}(U).$$

Obviously, g is R-linear and $g|N = f$; therefore $\overline{\mu} \in N$ and N = M follows. ■

<u>PROPOSITION</u> III.6.5. Let

$$\overline{0} \to E' \to E \to E'' \to \overline{0}$$

be an exact sequence in $\pi(R)$. Suppose that E is PK-injective and that E'' is PK-free, then E' is PK-injective.

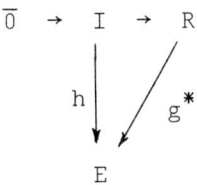

PROOF. Consider the following commutative diagram in $\pi(R)$, with exact rows :

$$\begin{array}{ccccccccc}
\overline{0} & \to & E' & \to & E & \to & E'' & \to & \overline{0} \\
 & & \uparrow{\scriptstyle f'} & & \uparrow{\scriptstyle f} & & \uparrow{\scriptstyle f''} & & \\
\overline{0} & \to & N & \to & M & \to & N/M & \to & \overline{0}
\end{array}$$

with M/N being PK-torsion and where, for given f' there is an f with $fj = if'$, while f" is the induced quotient map. Since $f''(M/N)$ is PK-torsion in E", it follows that $f''(M/N) = 0$, whence $f(M) \subset E'$. ■

PROPOSITION III.6.6. Let

$$\overline{0} \to E' \to E \to E'' \to \overline{0}$$

be an exact sequence in $\pi(R)$, where E' is PK-injective, E is PK-free and E" is PK-torsion, then $E = E'$.

PROOF. Since E' is PK-injective and PK-free, being a subobject of E; Proposition III.6.3. yields that E' is faithfully PK-injective. Hence the identity $1_{E'} : E' \to E'$ extends to a unique $\pi(R)$-morphism f making the following diagram commute :

$$\begin{array}{ccccccccc}
\overline{0} & \to & E' & \xrightarrow{i} & E & \to & E'' & \to & 0 \\
 & & & \searrow^{=} & \downarrow{\scriptstyle f} & & & & \\
 & & & & E' & & & &
\end{array}$$

Take $\overline{\mu} \in E$. If $E'' \neq \overline{0}$ then we may assume that $\overline{\mu} \neq \overline{0}$ since E" is flabby and $E(X) \to E''(X)$ is onto and non zero. Then $I\overline{\mu} \subset iE$ and $I\overline{\mu} \neq \overline{0}$ for some $I \in \mathcal{L}(K)$. It is clear that the Point $if(\overline{\mu}) - \overline{\mu}$ of E is annihilated by I, but as $PK(E) = \overline{0}$, if $(\overline{\mu}) = \overline{\mu}$ follows. This proves that $PE \subset iE'$ but since $E'' = PE''$ is the image of PE it follows that $E'' = \overline{0}$. Thus $E'' = \overline{0}$ and $E = E'$. ■

<u>PROPOSITION</u> III.6.7. Let $N' \subset N \subset M$ in $\pi(R)$ be such that M/N and N/N' are PK-torsion. Let $f : N \to E$ be given, where $E \in \pi(R)$ is PK-free. Suppose that the restriction $f|N'$ extends to $f^* : M \to E$, then f extends to f^*.

<u>PROOF</u>. Straightforward, mimicing the method of proof used for Proposition III.6.3. ■

<u>PROPOSITION</u> III.6.8. Let $M \in \pi(R)$ be K-torsion free, then M may be embedded in a faithfully PK-injective $E_{PK}(M) \in \pi(R)$ which is K-torsion free and such that $E_{PK}(M)/M$ is PK-torsion. $E_{PK}(M)$ is unique up to isomorphism.

<u>PROOF</u>. Let E be the injective hull $E^p(M)$ of M in $\pi(R)$ and consider the following exact sequence :

$$\overline{0} \to M \to E \xrightarrow{\pi} E/M \to \overline{0}$$

Define

$$E_{PK}(M) = E \underset{E/M}{\times} PK(E/M) \text{ (i.e. } \pi^{-1}(PK(E/M))).$$

Since $M \in F_K$ we get that $E \in F_K$ and also : $E_{PK}(M) \in F_K$ Moreover, $PK(E/M) \cong E_{PK}(M)/M$ since $M = E \underset{E/M}{\times} 0$, and thus $E_{PK}(M)/M$ is PK-torsion. From the exact sequence :

$$\overline{0} \to E_{PK}(M) \to E \to E/E_{PK}(M) \to \overline{0}$$

and Proposition III.6.5. we deduce that $E_{PK}(M)$ is PK-injective, hence faithfully PK-injective. Uniqueness up to isomorphism follows in the usual (universal) way. ■

If $M \in \pi(R)$ is arbitrary, define :

$$Q_{PK}(M) = E_{PK}(M/K(M));$$

this is called the <u>Pointwise Module of Quotients of M at K</u>. It follows directly from the construction that $Q_{PK}(M)$ equals $M/K(M) + PQ_{PK}(M)$. Unless $M/K(M)$ is flabby, $Q_{PK}(M)$ need not be flabby.

<u>LEMMA</u> III.6.9. Let $K \in$ R-Ker and let $M \in \pi(R)$ be a flabby presheaf of R-modules then $Q_{PK}(M) = PQ_K(M)$.

<u>PROOF</u>. The construction of $Q_K(M)$ in Section II.2. and the fact that the functor P is right semi-exact, imply that $PQ_K(M) \subset Q_{PK}(M)$. On the other hand, $M/K(M)$ is flabby, and we have an exact sequence in $\pi(R)$:

$$\overline{0} \to M/K(M) \to Q_{PK}(M) \to PK[E/(M/K(M))] \to \overline{0}$$

where E is the injective hull in $\pi(R)$ of $M/K(M)$. It follows that $Q_{PK}(M)$ is flabby, thus $Q_{PK}(M) \subset PQ_K(M)$. ■

<u>PROPOSITION</u> III.6.10. Q_{PK} is a left semi-exact endofunctor in $\pi(R)$.

<u>PROOF</u>. It is clear that Q_{PK} is a functor. Let $N,M \in \pi(R)$ and $N \subset M$. Denote $N/K(N)$, $M/K(M)$, by $\overline{N},\overline{M}$ resp. and let $E(\overline{N})$, $E(\overline{M})$ be injective hulls in $\pi(R)$ of $\overline{N},\overline{M}$ resp.. Since $\overline{0} \to N \to M$ is exact so are, $\overline{0} \to \overline{N} \to \overline{M}$ and $\overline{0} \to \overline{E}(N) \to \overline{E}(M)$. Because Q_K is left exact, the following commutative diagram has exact rows and columns :

$$\begin{array}{ccccccccc}
\overline{0} & \to & \overline{M} & \to & Q_K(N) & \xrightarrow{\pi_M} & K(E(\overline{M})/\overline{M}) & \to & \overline{0} \\
 & & \uparrow & & \uparrow{\scriptstyle i} & & \uparrow{\scriptstyle \pi} & & \\
\overline{0} & \to & \overline{N} & \to & Q_K(N) & \xrightarrow{\pi_N} & K(E(\overline{N})/\overline{N}) & \to & \overline{0} \\
 & & \uparrow & & \uparrow & & & & \\
 & & \overline{0} & & \overline{0} & & & &
\end{array}$$

By definition of $Q_{PK}(N)$, $Q_{PK}(N)$ maps to $PK(E(\overline{N})/\overline{N})$ under π_N, hence to $PK(E(\overline{M})/\overline{M})$ under $\pi \circ \pi_N$; therefore $iQ_{PK}(N) \subset Q_{PK}(M)$. ■

We focus on the Pointwise Ring of quotients in two cases :

Case 1. : R is a flabby Ring, $K \in$ R-Ker is local, but we do not impose here that K reduces R as in earlier sections.

Case 2. : R is a flabby Ring, $K \in$ R-Ker is contractible, i.e. $\mathcal{L}(K)$ is an idempotent filter and K has the following property : for $U \in$ Open (X) define $M|U \in \pi(R)$ by $(M|U)(V) = M(U \cap V)$ for all $V \in$ Open (X); K is contractible if $K(M|U) = K(M)|U$ for all $U \in$ Open (X).

The class of kernel functors which are both local and contractible is strongly linked to the structure of R.

PROPOSITION III.6.11. Let R be an arbitrary Ring, the following statements are equivalent for a local $K \in$ R-Ker :

1. K is contractible
2. $\overline{R}^W_V K(W) = K(V)$ for all $V \subset W$ in Open (X), i.e. K is known as soon as K(X) is given.

PROOF. 1. ⇒ 2. For $M \in \pi(R)$, $(K(M)|U)(W) = (K(M))(U \cap W) =$ $= K(U \cap W)'M(U \cap W))$ for every $U,W \in$ Open (X). On the other hand

$$K(M|U)(W) = K(W)((M|U)(W)) = K(W)(M(U\cap W)) = \overline{R}^W_{U\cap W}K(W)(M(U\cap W)).$$

Putting these results together, which is allowed because K is contractible, yields :

$$K(V)(M(V)) = \overline{R}^W_V K(W)(M(V)), \text{ writing } V \text{ for } U \cap W.$$

To an $R(V)$-module M_V there corresponds an $M \in \pi(R)$ such that $M(V) = M_V$, indeed take $M = C(M_V)$ as in Proposition III.2.3. Hence $K(V) = \overline{R}_V^W K(W)$ in $R(V)$-mod, follows.

2. $\Rightarrow$ 1. Since K is local, $L(K)$ is idempotent by Lemma III.6.2. Let $V \subset W$ in Open (X), then :

$$(K(M)|W)(V) = K(M)(V) = K(V)(M(V)) = \overline{R}_V^W K(W)(M(V)) = K(W)(M|V)) = K(M|W)(V).$$

■

<u>COROLLARY</u>. If R is the constant Ring then local contractible kernel functors K are constant, i.e. $K(U) = K(X)$ for all $U \in$ Open (X).

<u>THEOREM</u> III.6.12. (<u>Case 1</u>). <u>Let</u> R <u>be flabby and let</u> $K \in R$-Ker <u>be local</u>, <u>then</u> $Q_{PK}(R)$ <u>is a presheaf of rings</u>, <u>the Ring structure of</u> $Q_{PK}(R)$ <u>is uniquely determined by its</u> R-<u>Module structure</u>.

<u>PROOF</u>. $Q_{PK}(R)$ is a flabby presheaf because R is such. Therefore if $\mu \in Q_{PK}(R)$ there is a $\pi(R)$-morphism $\pi_\mu : R \to Q_{PK}(R)$ such that $\pi_\mu(X)1 = \pi \in Q_{PK}(R)(X)$, i.e. the Point $\overline{1} \in R$ maps to $\overline{\mu} \in Q_{PK}(R)$. There exists a unique $\pi(R)$-morphism $\hat{\pi}_\mu$ making the following diagram commute :

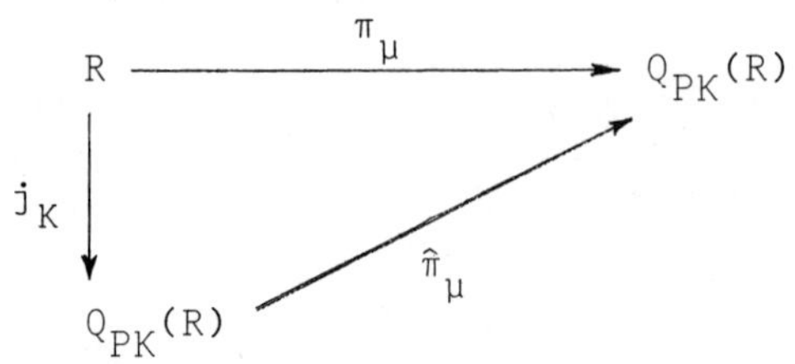

where j_K is exactly the morphism induced by $R \to R/K(R)$, hence $Q_{PK}(R)/j_K(R)$ is PK-torsion. If $\overline{\lambda} \in Q_{PK}(R)$ define $\overline{\lambda}.\overline{\mu} = \hat{\pi}_\mu(\overline{\lambda})$ by $(\overline{\lambda}.\overline{\mu})(U) = \hat{\pi}_\mu(U)(\overline{\lambda}(U))$. Let $I \in L(K)$ be such that $I\overline{\lambda} \subset R$, then :

$$\hat{\pi}_{\mu}(I\bar{\lambda}) = \mu_{\pi}(I\bar{\lambda}) = (I\bar{\lambda})\bar{\mu}.$$

On the other hand we also have :

$$\hat{\pi}_{\mu}(I\bar{\lambda}) = I\ \hat{\pi}_{\mu}(\bar{\lambda}) = I(\bar{\lambda}.\bar{\mu}).$$

Putting these together we see that the multiplication of Points in $Q_{PK}(R)$ just defined is compatible with the R-Module structure. Consider $\bar{\rho},\bar{\lambda},\bar{\mu} \in Q_{PK}(R)$. By definition $\bar{\rho}.(\bar{\lambda}.\bar{\mu}) = \hat{\pi}_{\eta}(\bar{\rho})$ where $\bar{\eta} = \bar{\lambda}.\bar{\mu}$. Choose $I \in L(K)$ such that $I\bar{\rho} \subset R/K(R)$ and $(I\bar{\rho})\bar{\lambda} \subset R/K(R)$. This is possible because $(I\bar{\rho})\bar{\lambda} = I(\bar{\rho}.\bar{\lambda})$; moreover we may assume that I is flabby since $L(K)$ is idempotent. Then :

$$I(\bar{\rho}.(\bar{\lambda}.\bar{\mu})) = \hat{\pi}_{\eta}(I\bar{\rho}) = (I\bar{\rho})(\bar{\lambda}.\bar{\mu}) = (I\bar{\rho})\hat{\pi}_{\mu}(\bar{\lambda})$$

$$= \hat{\pi}_{\mu}((I\bar{\rho})\bar{\lambda}) = ((I\bar{\rho})\bar{\lambda})\bar{\mu} = (I(\bar{\rho}.\bar{\lambda}))\bar{\mu} = I((\bar{\rho}.\bar{\lambda}).\bar{\mu}).$$

Since $Q_{PK}(R)$ is K-torsion free : $(\bar{\rho}.\bar{\lambda}).\bar{\mu} = \bar{\rho}.(\bar{\lambda}.\bar{\mu})$. Distributivities are easily verified.

Now, let $x,y \in Q_{PK}(R)(U)$ for some $U \in$ Open (X). Take $\bar{\nu},\bar{\mu} \in Q_{PK}(R)$ such that $\bar{\nu}(U) = x$, $\bar{\mu}(U) = y$ and define $xy = (\bar{\nu}.\bar{\mu})(U)$. If this is well-defined then a ring structure is defined in $Q_{PK}(R)(U)$ such that the restriction homomorphisms $Q_{PF}(R)^{U}_{V}$, $U \supset V$ in Open (X), are ring homomorphisms. The obtained Ring structure is clearly uniquely determined by the $\pi(R)$-structure of $Q_{PK}(R)$. In order to establish that the definition of xy is independent of the choice of Points $\bar{\nu},\bar{\mu}$ through x,y, it is not hard to see that it is sufficient to prove the following : if $\pi : R \to Q_{PK}(R)$ is a $\pi(R)$-morphism such that $\pi(U)$ is the zero map for some $U \in$ Open (X), and if $\hat{\pi} : Q_{PK}(R) \to Q_{PK}(R)$ is the extension of π, then $\hat{\pi}(U)$ is the zero map.

If $\bar{\mu} \in Q_{PK}(R)$ then $I\bar{\mu} \subset R$ for some $I \in L(K)$, hence

$$\hat{\pi}(U)(I(U)\bar{\mu}(U)) = \pi(U)(I(U)\bar{\mu}(U)) = 0$$

and $\hat{\pi}(U)(\overline{\mu}(U))$ is annihilated by $I(U) \in L(K(U))$. Since $Q_{PK}(R)$ is flabby,

$$\{\overline{\mu}(U),\ \overline{\mu} \in Q_{PK}(R)\} = Q_{PK}(R)(U),$$

hence $\hat{\pi}(U)(Q_{PK}(R)(U))$ is $K(U)$-torsion. However, $Q_{PK}(R)$ being K-torsion free, this yields that $\hat{\pi}(U) = 0$. ■

<u>PROPOSITION</u> III.6.13. Let $M \in \pi(R)$, $K \in R$-Ker local, then $PQ_K(M)$ is in a natural way a $Q_{PK}(R)$-Module.

<u>PROOF</u>. $Q_{PK}(PM)$ is flabby and by the left exactness of Q_K we get $Q_{PK}(PM) \subset PQ_K(M)$. Further, if $\overline{\mu} \in Q_K(M)$ then $\overline{I}\mu \subset M/K(M)$ for some flabby $I \in L(K)$, hence :

$$I\overline{\mu} \subset P(M/K(M)) = PM + K(M)/K(M) = \overline{PM}.$$

Right multiplication by $\overline{\mu}$ defines a $\pi(R)$-morphism

$$\pi : I \to \overline{PM} \to Q_{PK}(PM)$$

which extends to a $\pi(R)$-morphism $\hat{\pi}$; $R \to Q_{PK}(PM)$, because of Proposition III.6.4. The Point $\hat{\pi}(\overline{1})-\overline{\mu}$ of $Q_K(M)$ is annihilated by $I \in L(K)$, hence $\hat{\pi}(\overline{1}) = \overline{\mu}$ and $\overline{\mu} \in P_K(PM)$.
Thus $Q_{PK}(PM) = PQ_K(M)$ and we have reduced the problem to the case where M is flabby; then we proceed "exactly" as in the proof of the foregoing theorem to show that $Q_{PK}(M)$ (M then flabby) is a $Q_{PK}(R)$-Module with "scalar" multiplication induced by a "scalar" multiplication of Points of $Q_{PK}(M)$ by Points of $Q_{PK}(R)$. ■

<u>PROPOSITION</u> III.6.14. (Case 2.) Let R be flabby, $K \in R$-Ker contractible, then $Q_{PK}(R)$ is a Ring and its Ring structure is uniquely determined by its Module-structure.

PROOF. Repeat the proof of Theorem III.6.12 up to the point where we have to prove that $\pi(U) = 0$ implies $\hat{\pi}(U) = 0$. First we prove that if $M \in \pi(R)$, $I \in L(K)$ flabby, then for all $\bar{\mu} \in M$ we have $I(\bar{\mu}|U) = (I|U) = (I|U)(\bar{\mu}|U)$ for all $U \in \text{Open }(X)$, where $\bar{\mu}|U$ is defined by putting $(\bar{\mu}|U)(V) = \bar{\mu}(U \cap V)$, $V \in \text{Open }(X)$. Indeed, take $V \in \text{Open }(X)$, then $(I(\bar{\mu}|U))(V) = I(V)\bar{\mu}(U \cap V)$. By definition of the $\pi(R)$-structure of I the following holds :

$$I(V)\bar{\mu}(U \cap V) = R^{V}_{U \cap V}(I(V))\bar{\mu}(U \cap V).$$

Now, using the flabbiness of I, this implies :

$$(I(\bar{\mu}|U))(V) = I(U \cap V)\bar{\mu}(U \cap V) = ((I|U)(\bar{\mu}|U))(V).$$

With notations as before (Theorem III.6.12) we proceed as follows : If $\bar{\mu} \in Q_{PK}(R)$ then $I\bar{\mu} \subset R/K(R)$ for some flabby $I \in L(K)$. Since $\hat{\pi}(I\bar{\mu}) = \pi(I\bar{\mu})$, we have that $\hat{\pi}(I\bar{\mu})|U = \bar{0}$. Moreover :

$$(I\hat{\pi}(\bar{\mu}))|U = (I|U)(\hat{\pi}(\bar{\mu})|U) = I(\hat{\pi}(\mu)|U).$$

Hence $(I\hat{\pi}(\bar{\mu}))|U = \bar{0}$ or $\hat{\pi}(\bar{\mu})|U \in Q_{PK}(R)|U)$. Contractibility of K yields

$$K(Q_{PK}(R)|U) = K(Q_{PK}(R))|U = \bar{0}.$$

Therefore $\hat{\pi}(\bar{\mu})|U = \bar{0}$ for every $\bar{\mu} \in Q_{PK}(R)$, while $\hat{\pi}Q_{PK}(R)$ is flabby, hence $\hat{\pi}(U) = 0$. ■

PROPOSITION III.6.15. Let $K \in R\text{-Ker}$ be contractible then for every $M \in \pi(R)$, $PQ_K(M)$ is a $Q_{PK}(R)$-Module in a natural way.

PROOF. Similar to Proposition III.6.13. ■

Let I be a left (pre)-Ideal of R, then $Q_{PK}(PI) \subset Q_{PK}(I)$. By

Proposition III.6.13. or III.6.15. depending on the case. $PQ_K(I)$ is a $Q_{PK}(R)$-Module i.e. a left Ideal of $Q_{PK}(R)$. In general we have : $Q_{PK}(R)j_K(PI) \subset PQ_K(I)$. A kernel functor K is said to be a <u>T(P)-functor (or Pointwise T-functor)</u> if $Q_{PK}(R)j_K(PI) = PQ_K(I)$ for all $I \in L(K)$. Now

$$Q_K(I) = Q_K(R),\ PQ_K(I) = PQ_K(R) = Q_{PK}(R),$$

so K is a T(P)-functor if $Q_{PK}(R)j_K(I) = Q_{PK}(R)$ for all $I \in L(K)$.

<u>THEOREM</u> III.6.16. <u>Let</u> R <u>be a flabby Ring and suppose that</u> $K \in$ R-Ker <u>is either a local or contractible kernel functor</u>, <u>then</u> K <u>is a</u> T(P)-<u>functor if and only if for every left pre-Ideal</u> J <u>of</u> $Q_{PK}(R)$:

$$J = Q_{PK}(j_K^{-1}(J)) = Q_{PK}(R)J^c,$$

<u>where</u> $J^c = j_K j_K^{-1}(J) = J \cap j_K(R)$.

<u>PROOF</u>. Consider the commutative diagram, with exact rows and columns :

$$\begin{array}{ccccc} & & \bar{0} & & \bar{0} \\ & & \downarrow & & \downarrow \\ \bar{0} & \longrightarrow & J^c & \longrightarrow & J \\ & & \downarrow & & \downarrow \\ \bar{0} & \longrightarrow & Q_{PK}(J^c) & \longrightarrow & Q_{PK}(R) \end{array} \qquad (*)$$

As a subobject of $Q_{PK}(R)/j_K(R)$, J/J^c is K-torsion, while $Q_{PK}(J^c)/J^c$ is PK-torsion by construction.

<u>Case 1.</u> : <u>K is local</u>

J/J^c is a subpresheaf of $Q_{PK}(R)/j_K(R)$ and the $\pi(R)$-structure of both is defined via the morphism j_K. If $a \in (J/J^c)(U)$ then $a = \bar{\mu}(U)$ for some $\mu \in Q_{PK}(R)/j_K(R)$. Furthermore $I_{\bar{\mu}} = \bar{0}$ for some $I \in L(K)$. Denote $Q_{PK}(R)J^c$ by J^{ce}. If $b \in J(U)$ maps to a modulo

$J^c(U)$, then let $\overline{\nu} \in Q_{PK}(R)$ be such that $\overline{\nu}(U) = b$ and $\overline{\mu} = \overline{\nu} \bmod j_K(R)$. With $I \in L(K)$ chosen as above : $I\overline{\nu} \subset j_K(R)$ and $I(U)b \subset J^c(U)$. Hence

$$b \in (Q_{PK}(R)j_K(I))(U)b = (Q_{PK}(R))(U)b$$

and the latter is contained in $(Q_{PK}(R)J^c|U)$, in other words $J = Q_{PK}(R)J^c = J^{ce}$. On the other hand, $Q_{PK}(J^c)/J^c$ is PK-torsion, and the same holds for $Q_{PK}(J^c)/J^{ce}$. If $\overline{\mu} \in Q_{PK}(J^c)$ then $I\overline{\mu} \subset J^{ce}$ for some $I \in L(K)$ and as J^{ce} is a $Q_{PK}(R)$-Module : $Q_{PK}(R)I\overline{\mu} \subset J^{ce}$. Compatibility of the $\pi(R)$-structure with the Ring structure of $Q_{PK}(R)$ then entails that

$$Q_{PK}(R)I\overline{\mu} = Q_{PK}(R)j_K(I)\overline{\mu}.$$

Finally, since K is a T(P)-functor, $\overline{\mu} \in J^{ce}$, proving that $Q_{PK}(J^c) = J^{ce}$.

<u>Case 2.</u> : <u>K is contractible</u>

Put T equal to J or $Q_{PK}(J^c)$. Then T/J^c is K-torsion. Let $a \in T(U)$ and consider the diagram $(*)|U$ for $U \in$ Open (X). Obviously T/J^{ce} is K-torsion and thus $(T|U)/(J^{ce}|U)$ is K-torsion too. If $\overline{a} \in T|U$ then $I\overline{a} \subset J^{ce}|U$ for some flabby $I \in L(K)$. Since $J^{ce}|U$ is a $Q_{PK}(R)$-Module we have that $Q_{PK}(R)j_K(I)\overline{a} \subset J^{ce}|U$ and thus $\overline{a} \in J^{ce}|U$, i.e. $a \in J^{ce}$ for all $a \in T(U)$. So we may conclude that $T = J^{ce}$. For the converse : if $I \in L(K)$ then

$$Q_{PK}(PI) = PQ_K(I) = PQ_K(R) = Q_{PK}(R),$$

and $Q_{PK}(I) = Q_{PK}(R)$. If $J = Q_{PK}(R)j_K(I)$ then $J = Q_{PK}(j_K^{-1}(J))$, but $j_K^{-1}(J) \supset I$, whence $j_K^{-1}(J) \in L(K)$, $J = Q_{PK}(R)$. ■

THEOREM III.6.17. Let $K \in R\text{-Ker}$ be local, then the following statements are equivalent :

1. K is a T(P)-functor.

2. Every $M \in \pi(Q_{PK}(R))$ is PK-free.

PROOF. 1. ⇒ 2. Let $\bar{\mu} \in PK(M)$, then $I\bar{\mu} = \bar{0}$ for some $I \in L(K)$. This yields $Q_{PK}(R)j_K(I)\bar{\mu} = \bar{0}$ because PM is in a natural way a $Q_{PK}(R)$-Module and the $Q_{PK}(R)$-structure of M and PM is compatible with the R-structure of M and PM. By 1., $\bar{\mu} = \bar{0}$ follows.

2. ⇒ 1. Obviously :

$$Q_{PK}(R)j_K(PI) \subset Q_{PK}(j_K(PI)) = Q_{PK}(PI)$$

for every $I \in L(K)$. It follows that we have an inclusion :

$$Q_{PK}(PI)/Q_{PK}(R)j_K(PI) \to K(Q_{PK}(R)/Q_{PK}(R)j_K(PI)).$$

The smaller Module is flabby but also PK-free by 2., hence $Q_{PK}(PI) = Q_{PK}(R)j_K(PI)$. Left exactness of Q_K yields a monomorphism :

$$Q_K(R)/Q_K(PI) \to Q_K(R/PI)$$

Since $PI \in L(K)$, $Q_K(R/PI) = \bar{0}$ and $Q_K(R) = Q_K(PI)$, therefore taking Points yields

$$Q_{PK}(R) = PQ_K(PI) = Q_{PK}(PI) = Q_{PK}(R)j_K(PI). \blacksquare$$

THEOREM III.6.18. For a contractible kernel functor K, the following statements are equivalent :

1. K is a T(P)-functor.

2. Every $Q_{PK}(R)$-Module is PK-free.

3. Every $M \in \pi(Q_{PK}(R))$ is K-torsion free.

4. Every $M \in \pi(Q_{PK}(R))$ is PK-injective and K-torsion free.

PROOF. 1. $\Leftrightarrow$ 2. As before.

1. $\Rightarrow$ 3. Let $\overline{\mu} \in K(M)|U$, $U \in$ Open (X).

Contractibility of K implies that $I\overline{\mu} = \overline{0}$ for some $I \in L(K)$. Then $Q_{PK}(R)j_K(I)\overline{\mu} = \overline{0}$ and $\overline{\mu} = \overline{0}$ follows. Since to every non-zero $\mu \in K(M)(U)$ there corresponds a non-zero $\overline{\mu} \in K(M)|U$, it follows that $K(M)(U) = 0$ and this for all $U \in$ Open (X), hence $K(M) = \overline{0}$.

4. $\Rightarrow$ 3. Trivial; so is 4. $\Rightarrow$ 2.

1. $\Rightarrow$ 4. Since 1. implies 3. we only have to prove that 1. implies that every $M \in \pi(Q_{PK}(R))$ is PK-injective. Let $I \in L(K)$ and suppose that we are given a $\pi(R)$-morphism $f : I \to M$. Since M is K-torsion free, f induces a $\pi(R)$-morphism, again denoted by f,

$$f : I/K(I) \to M.$$

Note that $j_K(I) = I/K(I)$, where $j_K : R \to Q_K(R)$ is the canonical $\pi(R)$-morphism which is a Ring morphism when considered as a $\pi(R)$-morphism $R \to Q_{PK}(R)$. Since $Q_{PK}(R)j_K(I) = Q_{PK}(R)$ we have :

$$\overline{1} = \sum_i{}' \, \overline{q}_i\overline{a}_i, \text{ with } \overline{a}_i \in j_K(I),\ \overline{q}_i \in Q_{PK} \quad .$$

This follows from

$$1 = \sum_i{}' \, q_i a_i \text{ with } a_i \in j_K(I)(X),\ q_i \in Q_{PK}(R)(X).$$

Now, $I \cap (\bigcap_i (j_K(R) : \overline{q}_i)) \in L(K)$; let H be the flabby part of this presheaf. By idempotency of $L(K)$, $H \in L(K)$. Put $\overline{x} = \sum_i \overline{q}_i f(\overline{a}_i)$ and define $g : j_K(R) \to M$ by $g(\overline{1}) = \overline{x}$ and R-linearity. If $\overline{h} \in H$, then :

$$g(\overline{h}) = \overline{h}.\sum_i \overline{q}_i f(\overline{a}_i) = \sum_i (\overline{h}\overline{q}_i) f(\overline{a}_i)$$

$$= \sum_i f(\overline{h}\overline{q}_i\overline{a}_i) = f(\overline{h} \sum_i \overline{q}_i\overline{a}_i) = f(\overline{h}).$$

Therefore f and g coincide on $H \in L(K)$ and $g|I - f$ induces a $\pi(R)$-morphism $I/H \to M$ which has to be the zero map since $K(I/H) = I/H$ while $K(M) = \overline{0}$. Proposition III.6.4. finishes the proof. ■

REMARKS

1. One has to be careful when interpreting the multiplications of Points occuring in the above proof; these can be carried out because the Ring structure of $Q_{PK}(R)$ is compatible with the $\pi(R)$-structure. Where linearity with respect to multiplication by a Point has been used, this should be understood to be true "sectionwise".

2. The proof of the implication 1. ⇒ 4. may easily be adapted to yield a proof of : If K is a local T(P)-functor and $M \in \pi(Q_{PK}(R))$ is K-torsion free, then M is faithfully PK-injective.

3. If K is a local contractible kernel functor reducing R, then $Q_K(R)$ is easily seen to have injective restrictions maps, therefore $PQ_K(R)$ is a constant presheaf, hence $Q_{PK}(R)$ is a constant presheaf in that case.

4. If K takes flabby presheaves into flabby presheaves then it will follow from flabbiness of injective hulls that $Q_{PK} = Q_K$.

The reader is invited to investigate which properties still hold if one defines Pointwise localization in $\sigma(R)$ as follows $Q_{PK^S}(S) = \underline{\underline{a}}Q_{PK}(iS)$, where K is a $\sigma(R)$-compatible kernel functor in $\pi(R)$. Although this allows to study Pointwise Rings and Modules of quotients in $\sigma(R)$, one should not be too optimistic about the behaviour of Q_{PK^S}. The (non solved) problem is to find restrictive conditions on the topology of X, in order to have that sheafification of a flabby presheaf yields a flabby sheaf.

THEOREM III.6.19. Let R be a flabby ring, K a T(P)-functor which is either local or contractible then Q_{PK} is left and right semi-exact and commutes with direct sums.

PROOF. Case 1. : K is a local T(P)-functor. Left exactness of Q_{PK} has been established before. Let $M \xrightarrow{\pi} M' \to \bar{0}$ be an exact sequence in $\pi(R)$. In order to verify right semi-exactness of Q_{PK} we may assume that M and M' are K-torsion free. From the definitions, it follows that

$$Q_{PK}(M) = M + PQ_{PK}(M),\ Q_{PK}(M') = M' + PQ_{PK}(M').$$

It is clear that, to verify surjectivity of $\hat{\pi}$: $Q_{PK}(M) \to Q_{PK}(M')$ it will be sufficient to prove that $\hat{\pi}$ maps $PQ_{PK}(M)$ onto $PQ_{PK}(M')$. Take $\bar{\mu} \in Q_{PK}(M')$ then $I\bar{\mu} \subset M'$ for some flabby $I \in L(K)$. Put $N = P\pi^{-1}(I\bar{\mu}) \subset M$. Then $Q_{PK}(R)N \subset PQ_{PK}(M)$ since the latter is in a natural way a $Q_{PK}(R)$-Module, cf. Proposition III.6.13. Now $\hat{\pi}$ restricts to a $\pi(R)$-morphism $PQ_{PK}(M) \to PQ_{PK}(M')$ where PQ_{PK} (M') is a PK-free and a $Q_{PK}(R)$-Module. Let $\bar{\lambda} \in Q_{PK}(R) = PQ_K(R)$, then $I_1\bar{\lambda} \subset j_K(R)$ for some $I_1 \in L(K)$. Let $\bar{\nu} \in PQ_{PK}(M)$, then :

$$\hat{\pi}(I_1\bar{\lambda}.\bar{\nu}) = (I_1.\bar{\lambda})\hat{\pi}(\bar{\nu}) = I_1\hat{\pi}(\bar{\lambda}.\bar{\nu}),$$

hence :

$$\hat{\pi}(\bar{\lambda}.\bar{\eta}) - \bar{\eta}\hat{\pi}(\bar{\nu}) \in PKPQ_{PK}(M') = \bar{0}.$$

Therefore $\hat{\pi}$ restricts to a $\pi(Q_{PK}(R))$-morphism, whence :

$$\hat{\pi}(Q_{PK}(R)N) = Q_{PK}(R)\hat{\pi}(N) \supset Q_{PK}(R)(I\bar{\mu})$$

and by definition of the $Q_{PK}(R)$-structure and by the T(P)-

condition , it follows that $\bar{\pi}(Q_{PK}(R)N) \supset Q_{PK}(R)\bar{\mu}$. Thus there exists a Point of $Q_{PK}(R)N$ which maps to $\bar{\mu}$ under $\bar{\pi}$. We know already that

$$PQ_{PK}(\bigoplus_i M_i) = Q_{PK}(P(\bigoplus_i M_i)) = Q_{PK}(\bigoplus_i PM_i).$$

In the exact sequence :

$$\overline{0} \to \bigoplus_i PM_i \to \bigoplus_i Q_{PK}(PM_i) \to \bigoplus_i (Q_{PK}(PM_i)/PM_i) \to \overline{0}$$

$\bigoplus_i (Q_{PK}(PM_i)/PM_i)$ is PK-torsion, while $\bigoplus_i Q_{PK}(PM_i)$ is K-torsion free and a $Q_{PK}(R)$-Module. We may now repeat the proof of Theorem III.6.18, 3. ⇒ 4., for this particular $M = \bigoplus_i Q_{PK}(PM_i)$ which is now K-torsion free by construction, (while Proposition III.6.4. which is being used in the proof does not discriminate between local and contractible) to obtain that $\bigoplus_i Q_{PK}(PM_i)$ is faithfully PK-injective. So we proved up to now that $\bigoplus_i Q_{PK}(PM_i) \cong Q_{PK}(\bigoplus PM_i)$. It is easily seen that for all $M \in \pi(R)$, $PQ_{PK}(M) = Q_{PK}(PM)$, hence $\bigoplus_i Q_{PK}(PM_i) = \bigoplus_i PQ_{PK}(M_i)$ but also

$$\bigoplus_i Q_{PK}(PM_i) \cong Q_{PK}(P(\bigoplus_i M_i)) \cong PQ_{PK}(\bigoplus_i M_i).$$

Since

$$Q_{PK}(\bigoplus_i M_i) = \bigoplus_i \overline{M}_i + PQ_{PK}(\bigoplus M_i),$$

the sum being taken in $Q_K(\bigoplus_i M_i)$, and since $\oplus$ is compatible with this sum, one easily derives that

$$Q_{PK}(\bigoplus_i M_i) = \bigoplus_i (\overline{M}_i + PQ_{PK}(M_i)) = \bigoplus_i Q_{PK}(M_i),$$

where $\overline{M}_i$ denotes $M_i/K(M_i)$.

Note that the same argumentation works if K were contractible so we also proved that Q_{PK} commutes with direct sums if K is contractible (although there is an easier proof in that case).

Case 2. : K is a contractible T(P)-functor. We only have to prove right semi-exactness of Q_{PK}. This follows as in Case 1 but substituting Proposition III.6.15. for III.6.13. ■

III.7. Strictly Local Kernel Functors and Coherent Modules

III.7.1. Old Hat on Quasi-Coherence and Coherence

Although many of the results in this Chapter hold for presheaves, we will restrict attention to sheaves. A topological space X together with a sheaf of rings R will be referred to as the ringed space (X,R). Let M be an R-Module. M is said to be quasi-coherent if each point x has an open neighborhood U such that the restriction $M|U$ of M to U is isomorphic to the cokernel of a morphism $R^{(I)}|U \to R^{(J)}|U$, where I and J are arbitrary index sets. Clearly, R itself is quasi-coherent and each finite sum of quasi-coherent Modules is quasi-coherent. A sheaf of R-Modules M is said to be coherent if it is of finite type and if for each open U in X, each $n \geq 0$ and each morphism $u : R^n|U \to M|U$, the kernel of u is of finite type. The main fact about coherent Modules is the following : if in a short exact sequence in $\sigma(R)$, two out of three Modules are coherent, then so is the third Module.
Let QC(X,R) be the category of quasi-coherent R-Modules and let C(X,R) be the category of coherent R-Modules.

LEMMA III.7.1. QC(X,R) is a Grothendieck category, with generator R.
Note that C(X,R) does not have to form a Grothendieck category as e.g. each quasi-coherent R-Module is the inductive limit of its coherent subModules, because each R-Module is inductive limit of its sub-R-Modules of finite type.

PROPOSITION III.7.2. Let R be a coherent sheaf of rings on the topological space X, and M a coherent sheaf of R-modules. Let $x \in X$, and assume that N^x is a submodule of finite type of the stalk M_x, then, on a sufficiently small neighborhood of x there is a coherent subModule N of M such that $N_x = N^x$.

PROOF. Let us recall the proof as it is given in [13]. Elements of M(U) may be viewed as sections of M over U, i.e. continuous functions $f : U \to \tilde{M}$ where $(\tilde{M}, X, \pi)$ is the sheaf space attached to M, $\pi \circ f = 1_U$. Assume that $\{n_1, \ldots, n_k\}$ generates N^x as an R_x-module. Since, for each i, we have $n_i \in M_x$, we get that $n_i = f_i(x)$ where f_i is a section of M over some open neighborhood of x. Since there are only a finite number of f_i we may assume that all f_i are defined over the same open neighborhood U of x. It follows that we have a homeomorphism $f : R^k|U \to M|U$. Take $N = \mathrm{Im}\, f$, then N is coherent since it is a subModule of finite type of a coherent sheaf; clearly $N_x = N^x$. ■

PROPOSITION III.7.3. Let K be a local kernel functor reducing R and let K be a T-functor, then Q_K is inner in QC(R).

PROOF. Immediate from Theorem III.5.1. 2.

III.7.2. Strictly Local Kernel Functors

A kernel functor K in $\sigma(R)$ is said to be local at $x \in X$ if : for all $M, M' \in \sigma(R)$, $M_x = M'_x$ implies $(KM)_x = (KM')_x$, where M_x, M'_x are the stalks of M,M' resp. at x. K is said to be a strictly local kernel functor if it is local at x for every $x \in X$.

PROPOSITION III.7.4. Let K be a kernel functor in $\sigma(R)$ which is local at $x \in X$, then there is a (unique) kernel functor K_x in R_x-mod, such that $K_x(M_x) = (KM)_x$ for every $M \in \sigma(R)$.

PROOF. Let M^x be any R_x-Module. Define $C(M^x) \in \pi(R)$ by $C(M^x)(U) = 0$ if $x \notin U$, $C(M^x)(U) = (R_x^U)_*(M^x)$ if $x \in U$. One easily checks that $C(M^x) \in \sigma(R)$ and that $C(M^x)_x = M^x$. Define K_x by $K_x(M^x) = (K(C(M^x)))_x$. Since K is local at x, if M is any sheaf such that $M_x = M^x$ then $(K(M))_x = K_x(M^x)$. Now, K_x is a subfunctor of the identity in R_x-mod, which is left exact because C, K and taking inductive limits, i.e. taking stalks, are left exact functors. Furthermore :

$$M^x/K_x(M^x) = M^x/(K(C(M^x)))_x =$$

$$(C(M^x))_x/(K(C(M^x)))_x = (C(M^x)/K(C(M^x)))_x.$$

Hence, again using the fact that K is local at x :

$$K_x(M^x/K_x(M^x)) = (K(C(M^x)/K(C(M^x)))_x = \overline{0}_x = 0. \blacksquare$$

PROPOSITION III.7.5. Let R be a coherent sheaf of left Noetherian rings, let E be injective in $\sigma(R)$. Then, for every $x \in X$, E_x is an injective R_x-module.

PROOF. We have to show that each diagram

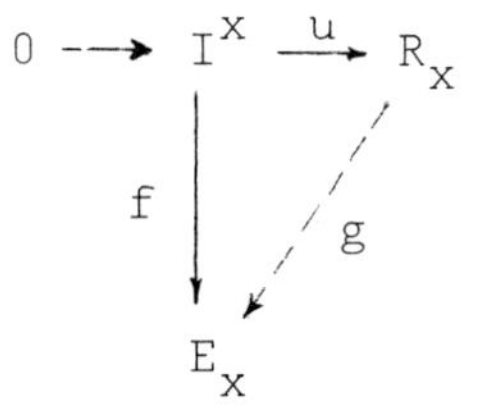

where I^x is a left ideal of R_x, $f : I^x \to E_x$ an R_x-linear map, may be completed by an R_x-linear g to make it into a commutative diagram i.e. gu = f. It is not hard to verify that the restriction of E to an open set is injective. So to prove the statement it suffices to find a neighborhood U of x, a left Ideal I of R|U on U, such that $I_x = I^x$, and a morphism $v : I \to R|U$ inducing u in I^x. All this follows from Proposition III.7.2. Injectivity of E|U finishes the proof. ■

Note : The above proposition has been mentioned in [13].

PROPOSITION III.7.6. Let R be a coherent sheaf of left Noetherian rings and let M be a sheaf of left R-modules, then $E^s(M)_x = E_x(M_x)$, where $E_x(M_x)$ is the injective hull of M_x in R_x-mod.

PROOF. From the inclusion of M in $E^s(M)$, we get $M_x \subset E^s(M)_x$. But then $E_x(M_x) \subset E^s(M)_x$, since the latter is injective in R_x-mod by the foregoing proposition. We have $M \subset E^0(M) \subset E^s(M)$, where $E^0(M)$ is the sheaf defined by $E^0(M)(U) = \prod_{x \in U} E_x(M_x)$ for U ∈ Open (X). It is easily verified that $E^0(M)$ is injective in σ(R), therefore $E^0(M) = E^s(M)$, hence $E^s(M)_x = E_x(M_x)$. ■

Recall that, if K is a kernel functor in σ(R) and if M ∈ σ(R) is K-torsion free, then the K-injective hull of M is defined to be

$$E_K(M) = M \underset{E^s(M)/M}{\times} K(E^s(M)/M);$$

PROPOSITION III.7.7. Let R be a coherent sheaf of left Noetherian rings, let K be a kernel functor in σ(R) which is local at $x \in X$. If $M \in \sigma(R)$, then :

$$(Q_K(M))_x = Q_{K_x}(M_x).$$

PROOF. Assume first that M is K-torsion free, i.e. $Q_K(M) = E_K(M)$ by definition. Then :

$$(Q_K(M))_x = (E_K(M))_x = (M \underset{E^s(M)/M}{\times} K(E^s(M)/M))_x$$

$$= M_x \underset{(E^s(M)/M)_x}{\times} (K(E^s(M)/M))_x$$

and as K is local at $x \in X$, we get

$$(Q_K(M))_x = M_x \underset{E^s(M)_x/M_x}{\times} K_x(E^s(M)_x/M_x)$$

$$= M_x \underset{E_x(M_x)/M_x}{\times} (E_x(M_x)/M_x) = E_{K_x}(M_x).$$

The fact that M is K-torsion free implies that M_x is K_x-torsion free, whence : $(Q_K(M))_x = Q_{K_x}(M_x)$. For arbitrary $M \in \sigma(R)$ we obtain :

$$(Q_K(M))_x = (E_K(M/K(M)))_x = E_{K_x}(M_x(K_x(M_x)) = Q_{K_x}(M_x). \blacksquare$$

PROPOSITION III.7.8. Let K be a strictly local kernel functor in $\sigma(R)$, where R is a coherent sheaf of left Noetherian rings, then $Q_K(R)$ is a sheaf of rings. If $M \in \sigma(R)$ then $Q_K(M)$ is in a natural way a $Q_K(R)$-Module.

PROOF. Immediate from the foregoing and the corresponding facts for rings and modules. ■

THEOREM III.7.9. Let R be a coherent sheaf of left Noetherian rings. A strictly local kernel functor K in $\sigma(R)$ is a T-functor

<u>if and only if</u>, <u>for each</u> $x \in X$, <u>the kernel functor</u> K_x <u>in</u> R_x-mod <u>is a t-functor</u>.

<u>PROOF</u>. Suppose K is a T-functor and take $M^x \in Q_{K_x}(R_x)$-mod. Since $(Q_K(R))_x = Q_{K_x}(R_x)$, it follows that the "canonical" sheaf $C(M^x)$ associated to M^x in $\sigma(R)$, actually has a $Q_K(R)$-Module structure, therefore it is K-torsion free. Because K is strictly local : $K_x(M^x) = (K(C(M^x)))_x = 0$, hence $K_x(M^x) = 0$ for every $M^x \in R_x$-mod, what means that K_x is a t-functor. The converse is even more obvious. ■

<u>COROLLARY</u>. Let K be a strictly local T-functor and let R be a coherent sheaf of left Noetherian rings, then $Q_K(R)$ is a coherent sheaf of left Noetherian rings. If $M \in \sigma(R)$ is coherent then $Q_K(M)$ is a coherent $Q_K(R)$-Module.

III.7.3. <u>Some Examples</u>

<u>PROPOSITION</u> III.7.10. Let K be a local kernel functor in $\sigma(R)$. Let $x \in X$ and suppose that x has a fundamental set of neighborhoods, $V(x)$, such that for each $U \in V(x)$ the kernel functor K(U) is Noetherian; then K is local at x and

$$K_x = \sup_{U \ni x} \bar{R}^U_x(K(U)),$$

where :

$$\bar{R}^U_x(K(U))(M^x) = (R^U_x)^*[K(U)(R^U_x)_*(M^x))].$$

<u>PROOF</u>. It will suffice to check that, for all $M \in \sigma(R)$, we have

$$[\sup_{U \ni x} \bar{R}^U_x(K(U))](M_x) = \varinjlim_{U \ni x} K(U)(M(U)) = (K(M))_x.$$

First if $V \in \mathcal{V}(x)$ then one easily verifies that $\bar{R}_x^V K(V)$ is a Noetherian kernel functor, hence it commutes with direct sums and direct limits. Let $\{K_i, i \in I\}$ be an inductive system of kernel functors in R_x-mod such that for every i there exists $j \geq i$ such that K_j is Noetherian then

$$[\sup_{i \in I} K_i](M^x) = \varinjlim_{i \in I} K_i(M^x), \text{ for every } M^x \in R_x\text{-mod}.$$

Left exactness of $\varinjlim$ yields that the right hand side is a left exact endofunctor in R_x-mod. Moreover, right exactness of $\varinjlim$ yields that

$$M^x/\varinjlim_{i \in I} K_i(M^x) = \varinjlim_{i \in I} (M^x/K_i(M^x))$$

and $$\varinjlim_{i \in I} K_i(M^x/\varinjlim_{i \in I} K_i(M^x)) = \varinjlim_{j \in J} K_j(\varinjlim_{i \in I} (M^x)K_i(M^x)))$$

where J indexes the cofinal set of Noetherian K_i. Hence, the latter R_x-module equals

$$\varinjlim_{j \in J} \varinjlim_{i \in I} K_j(M^x/K_i(M^x)) = \varinjlim_{j \in J} K_j(M^x)) = 0.$$

This proves that putting $K(M^x) = \varinjlim_{i \in I} K_i(M^x)$ defines the smallest idempotent kernel functor in R_x-mod which is bigger than K_i for all $i \in I$, i.e. $K = \sup_{i \in I} K_i$.

If $U \supset V$ in Open (X), then the fact that K is local yields $\bar{R}_V^U K(U) \leq K(V)$, hence

$$\bar{R}_x^V \bar{R}_V^U K(U) = \bar{R}_x^U K(U) < \bar{R}_x^V K(V).$$

So the foregoing applies to the system

$$\{\bar{R}_x^V K(V), V \text{ in Open } (X)\}.$$

For R(U)-modules, the R(U)-structure of which is defined by $(R_x^U)_*$, we will not write the $(R_x^U)^*$ if we mean to consider them as R_x-modules because they are R_x-modules in the obvious way. Now :

$$\varinjlim_{U \ni x} K(U)[(R_x^U)_*(M_x)] = \varinjlim_{U \ni x} (\bar{R}_x^U K(U))(M_x),$$

by definition of the lattice morphism $\bar{R}_x^U$. Applying our foregoing argumentation, we obtain :

$$\varinjlim_{U \ni x} (\bar{R}_x^U K(U))(M_x) = (\sup_{U \ni x} \bar{R}_x^U K(U))(M_x).$$

On the other hand we have :

$$\varinjlim_{U \ni x} K(U)[(R_x^U)_*(M_x)] = \varinjlim_{U \ni x} K(U)[(R_x^U)_* \varinjlim_{\substack{V \in \gamma(x) \\ U \supset V}} M(V)]$$

$$= \varinjlim_{U \in \gamma(x)} K(U)[(R_x^U)_* \varinjlim_{\substack{V \in \gamma(x) \\ U \supset V}} M(V)]$$

and because $\gamma(x)$ is cofinal in the set of opens containing x,

$$= \varinjlim_{U \in \gamma(x)} K(U)[\varinjlim_{\substack{V \in \gamma(x) \\ U \supset V}} (R_V^U)_* M(V)]$$

and now, since $U \in \gamma(x)$ yields that K(U) commutes with direct limits we get :

$$= \varinjlim_{V \in \gamma(x)} \varinjlim_{\substack{V \in \gamma(x) \\ U \supset V}} K(U)(R_V^U)_* M(V)$$

however, $K(U)(R_V^U)_* M(V) \leq K(V)M(V)$ and every V appearing in the second $\varinjlim$ also appears in the first, so :

$$= \varinjlim_{V \in \gamma(x)} K(V)M(V) = (K(M))_x$$

as desired. ■

<u>COROLLARY</u>. If X has a basis $\mathcal{B}$ such that for all $B \in \mathcal{B}$ the kernel functor $K(B)$ in $R(B)$-mod is Noetherian, then K is strictly local and for all x in X, K_x in R_x-mod is given by $\sup\limits_{U \ni x} \overline{R}_x^U K(U)$. Together with Proposition III.7.7., this yields a strengthening of Theorem I.3.4. and Proposition I.3.7. However one should bear in mind that in this case K is presupposed to be inner in $\sigma(R)$!

Recall that a morphism $a \to b$ in a category $\underline{C}$ is an epimorphism in $\underline{C}$ if the following diagram is a pushout diagram :

$$\begin{array}{ccc} a & \longrightarrow & b \\ \downarrow & & \downarrow 1 \\ b & \xrightarrow[1]{} & b \end{array}$$

If $\underline{C}$ is the category of rings this means that $R_1 \to R_2$ is epimorphic exactly then when $R_2 \cong R_2 \underset{R_1}{\otimes} R_2$. If $\underline{C}$ is the category of sheaves of rings, then a morphism is an Epimorphism if $R_2 \cong R_2 \underset{R_1}{\otimes} R_2$ and this is easily seen to be equivalent to : for all $x \in X$, $(R_1)_x \to (R_2)_x$ is an epimorphism in the category of rings.
Let $f : R \to F$ be a Ring-Epimorphism and suppose that F is a flat right R-Module, where the right R-Module structure of F is defined via f. For $M \in \sigma(R)$, denote by $K_{(F)}(M)$ the kernel of the canonical morphism $\psi_M : M \to F \underset{R}{\otimes} M$, then :

<u>PROPOSITION</u> III.7.11. $K_{(F)}$ is a strictly local kernel functor in $\sigma(R)$.

<u>PROOF</u>. It is obvious that $K_{(F)}$ is a subfunctor of the identity in $\sigma(R)$. Let

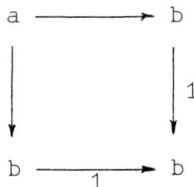

$$\overline{0} \to M' \to M \to M'' \to \overline{0}$$

be an exact sequence in $\sigma(R)$. Flatness of f yields the following exact commutative diagram in $\sigma(R)$:

$$\begin{array}{ccccccccc}
 & & \overline{0} & & \overline{0} & & \overline{0} & & \\
 & & \downarrow & & \downarrow & & \downarrow & & \\
\overline{0} & \longrightarrow & K_{(F)}(M') & \longrightarrow & K_{(F)}(M) & \longrightarrow & K_{(F)}(M'') & & \\
 & & \downarrow & & \downarrow & & \downarrow & & \\
\overline{0} & \longrightarrow & M' & \longrightarrow & M & \longrightarrow & M'' & \longrightarrow & \overline{0} \\
 & & \downarrow & & \downarrow & & \downarrow & & \\
\overline{0} & \longrightarrow & F \underset{R}{\otimes} M' & \longrightarrow & F \underset{R}{\otimes} M & \longrightarrow & F \underset{R}{\otimes} M'' & \longrightarrow & 0
\end{array}$$

proving that $K_{(F)}$ is left exact.

Now $N = K_{(F)}\ (K_{(F)}(M))$ is defined by the exact sequence :

$$\overline{0} \to N \to K_{(F)}(M) \to F \underset{R}{\otimes} K_{(F)}(M)$$

and by right flatness of f we obtain an exact sequence :

$$\overline{0} \longrightarrow F \underset{R}{\otimes} K_{(F)}(M) \longrightarrow F \underset{R}{\otimes} M \xrightarrow{\pi} F \underset{R}{\otimes} (F \underset{R}{\otimes} M)$$

However, since f is an Epimorphism we have that $F \underset{R}{\otimes} F$ is isomorphic to F as an R-biModule, therefore π is an isomorphism, hence $F \underset{R}{\otimes} K_{(F)}(M) = \overline{0}$. Then it follows from the first exact sequence that $N = K_{(F)}(M)$, i.e. $K^2_{(F)}(M) = K_{(F)}(M)$. For $M \in \sigma(R)$, consider the exact commutative diagram :

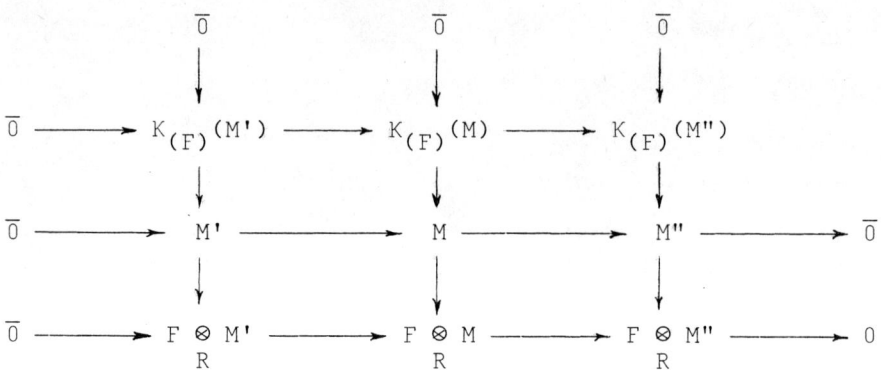

$$
\begin{array}{ccccccccc}
 & & \overline{0} & & \overline{0} & & \overline{0} & & \\
 & & \downarrow & & \downarrow & & \downarrow & & \\
\overline{0} & \longrightarrow & K_{(F)}(K_{(F)}(M)) & \xrightarrow{i} & K_{(F)}(M) & \longrightarrow & K_{(F)}(M/K_{(F)}(M)) & & \\
 & & i\downarrow & & \downarrow & & \downarrow & & \\
\overline{0} & \longrightarrow & K_{(F)}(M) & \longrightarrow & M & \longrightarrow & M/K_{(F)}(M) & \longrightarrow & \overline{0} \\
 & & \psi\downarrow & & \downarrow & & \downarrow & & \\
\overline{0} & \longrightarrow & F \underset{R}{\otimes} K_{(F)}(M) & \longrightarrow & F \underset{R}{\otimes} M & \longrightarrow & F \underset{R}{\otimes} (M/K_{(F)}(M)) & \longrightarrow & \overline{0} \\
 & & \downarrow & & & & & & \\
 & & \text{Coker } \psi & & & & & & \\
 & & \downarrow & & & & & & \\
 & & \overline{0} & & & & & &
\end{array}
$$

Applying the snake lemma yields an exact sequence

$$\overline{0} \to K_{(F)}(K_{(F)}(M)) \xrightarrow{i} K_{(F)}(M) \to K_{(F)}(M/K_{(F)}(M)) \to \text{Coker } \psi \to \ldots$$

Hence, since i is an isomorphism and since $F \underset{R}{\otimes} K_{(F)}(M) = \overline{0}$ implies Coker $\psi = \overline{0}$, the conclusion is that $K_{(F)}(M/K_{(F)}(M)) = \overline{0}$, i.e. $K_{(F)}$ is a kernel functor in $\sigma(R)$. Furthermore :

$$
\begin{aligned}
(K_{(F)}(M))_x &= (\text{Ker}(M \to F \underset{R}{\otimes} M))_x \\
&= \text{Ker}(M_x \to (F \underset{R}{\otimes} M)_x) \\
&= \text{Ker}(M_x \to F_x \underset{R_x}{\otimes} M_x) = \kappa_{F_x}(M_x)
\end{aligned}
$$

where κ_{F_x} is the kernel functor in R_x-mod defined by Γ_x. ■

<u>PROPOSITION</u> III.7.12. The kernel functor $K_{(F)}$, where F is as above, is a T-functor; moreover we have that $Q_{K_{(F)}}(R) \cong F$, i.e. $Q_{K_{(F)}}(M) \cong F \underset{R}{\otimes} M$ for all $M \in \sigma(R)$.

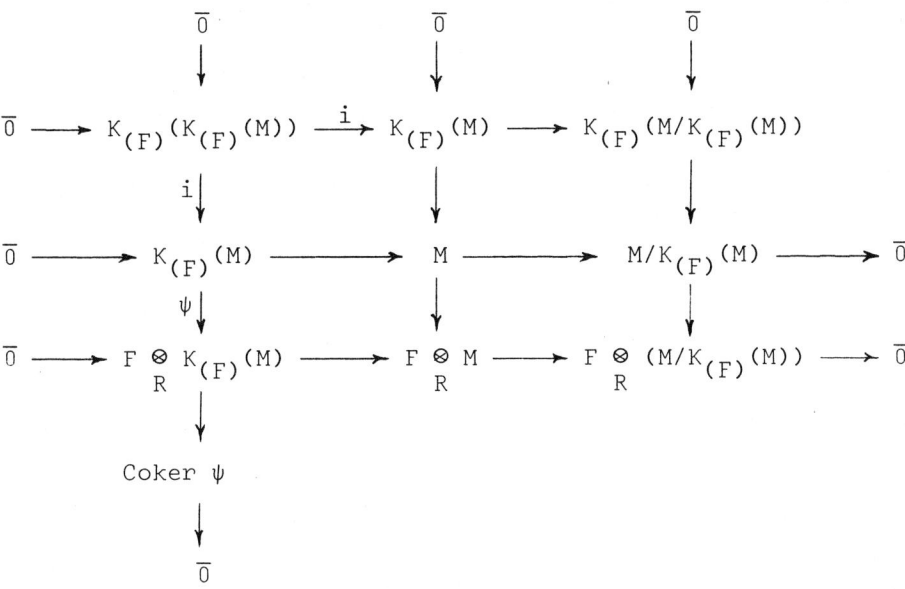

PROOF. Let $M \in \sigma(R)$ be $K_{(F)}$-torsion free, then

$$\overline{0} \to M \to F \underset{R}{\otimes} M$$

is exact. It is trivial that $F \underset{R}{\otimes} M$ is $K_{(F)}$-torsion free and moreover $F \underset{R}{\otimes} M/M$ is $K_{(F)}$-torsion since

$$F \underset{R}{\otimes} (F \underset{R}{\otimes} M/M) = (F \underset{R}{\otimes} F) \underset{R}{\otimes} M/F \otimes M = \overline{0}$$

We only have to prove that $F \underset{R}{\otimes}$ is $K_{(F)}$-injective. Let there be given a diagram of $\sigma(R)$-morphisms

$$\begin{array}{l} \overline{0} \\ \downarrow \\ N' \xrightarrow{f} F \underset{R}{\otimes} M \\ \downarrow \\ N \\ \downarrow \\ N/N' \\ \downarrow \\ \overline{0} \end{array}$$

where N/N' is $K_{(F)}$-torsion, i.e. $F \underset{R}{\otimes} (N/N') = \overline{0}$. Tensoring on the right by F yields :

$$\begin{array}{l} \overline{0} \\ \downarrow \\ F \underset{R}{\otimes} N' \xrightarrow{\quad 1_F \otimes f \quad} F \underset{R}{\otimes} (F \underset{R}{\otimes} M) \cong F \underset{R}{\otimes} M \\ \downarrow \\ F \underset{R}{\otimes} N \\ \downarrow \\ \overline{0} \end{array}$$

The isomorphism $F \underset{R}{\otimes} N \to F \underset{R}{\otimes} N'$ combines with $1_P \otimes f$ to yield a

$F \otimes_R N \to F \otimes_R M$ which yields a morphism $g : N \to F \otimes_R M$, defined to be the composed morphism : $N \to F \otimes_R N \to F \otimes_R M$, and this obviously extends f. Uniqueness of $Q_{K_{(F)}}(M)$ with the above properties, proves $Q_{K_{(F)}}(M) \cong F \otimes_R M$.

In general for arbitrary M we have an exact commutative diagram :

$$\begin{array}{ccccccccc} \overline{0} & \longrightarrow & K_{(F)}(M) & \longrightarrow & M & \longrightarrow & M/K_{(F)}(M) & \longrightarrow & \overline{0} \\ & & \downarrow & & \downarrow & & \downarrow & & \\ \overline{0} & \longrightarrow & F \otimes_R K_{(F)}(M) & \longrightarrow & F \otimes_R M & \longrightarrow & F \otimes_R (M/K_{(F)}(M)) & \to & \overline{0} \end{array}$$

Since $F \otimes_R K_{(F)}(M) = \overline{0}$ it follows that :

$$F \otimes_R M \cong F \otimes_R (M/K_{(F)}(M)) \cong Q_{K_{(F)}}(M).$$

Taking M = F yields $Q_{K_{(F)}}(R) \cong F$ and the fact that $K_{(F)}$ is a T-functor is then a trivial consequence of :

$$(Q_{K_{(F)}}(M))_x = Q_{(K_{(F)})_x}(M_x) = Q_{\kappa_{F_x}}(M_x) = F_x \otimes_{R_x} M_x,$$

where $R_x \to F_x$ is a right flat ring epimorphism, i.e. κ_{F_x} is a t-functor for all $x \in X$. ■

Let us conclude with the following example : let A be a left Noetherian ring with unit and let R be the constant sheaf over the topological space X = spec A with stalk A. Give a local kernel functor K in $\pi(R)$ associating Q_I to the Zariski open set X_I, I an ideal of A, as in Section I.3. Suppose that K is a T-functor, e.g. if A is commutative or Zariski central. Since R is a coherent sheaf of left Noetherian ring it follows that K is strictly local. Hence Spec may be considered as an exact functor in $\pi(R)$ which commutes with direct sums and which is strictly local, hence it is given by a bunch of t-functors $\{K_x, x \in X\}$. If $M \in \pi(R)$, e.g. $M = \tilde{m}$ for some $m \in A$-mod, then $(Q_K(M))_x = Q_{K_x}(M_x)$ for all $x \in X$, proving that in Proposition I.3.5. the hypothesis of M being of finite type may be dropped.

CHAPTER IV. APPLICATIONS

In this chapter we exhibit some geometrical applications of the theory we derived.

IV.1. Kerneled Spaces

(Pre) Structured spaces and morphisms of (pre) structured spaces, e.g. ringed spaces, are defined as usual. We define a (pre) kerneled space to be a triple (X,R,K) where : X is a topological space, R is a (pre) sheaf of rings such that (X,R) is a ringed space and K is a kernel functor on $(\pi(R))$, $\sigma(R)$. The class of all (pre) kerneled spaces, denoted by (K^p), K^s, may be given the structure of a category as follows.

Let $S_1 = (X_1,R_1,K_1)$ and $S_2 = (X_2,R_2,K_2)$ be (pre) kerneled spaces, then $\mathrm{Hom}_K(S_1,S_2)$ consists of all couples (f,Θ) where (f,Θ) is a morphism of (pre) ringed spaces,

$$(f,\Theta) : (X_1,R_1) \to (X_2,R_2),$$

satisfying the following condition : if $M_1 \in \sigma(R_1)$ (resp. $\pi(R_1)$), $M_2 \in \sigma(R_2)$ (resp. $\pi(R_2)$), and if (f,ψ) is a morphism of (pre) structured spaces which is semilinear with respect to (f,Θ), then we get a commutative diagram of morphisms of (pre) structured spaces :

$$\begin{array}{ccc}
(X_1,M_1) & \xrightarrow{(f,\psi)} & (X_2,M_2) \\
\uparrow{\scriptstyle (1_{X_1},i_1)} & & \uparrow{\scriptstyle (1_{X_2},i_2)} \\
(X_1,K_1(M_1)) & \xrightarrow[(f,\psi_{res})]{} & (X_2,K_2(M_2))
\end{array}$$

where i_ν, $\nu = 1,2$, is the canonical inclusion of $K_\nu(M_\nu)$ in M_ν and where ψ_{res} is the restriction of ψ to $K_2(M_2)$.
Equivalently, for each $U \in \text{Open}\ (X_2)$ we get :

$$\begin{array}{ccc}
M_2(U) & \xrightarrow{\psi(U)} & M_1(f^{-1}(U)) \\
\uparrow{\scriptstyle i_2(U)} & & \uparrow{\scriptstyle i_1(f^{-1}(U))} \\
K_2(M_2)(U) & \xrightarrow[\psi_{res}(U)]{} & K_1(M_1)(f^{-1}(U))
\end{array}$$

<u>LEMMA</u> IV.1. Consider a morphism of (pre) kerneled spaces :

$$(f,\Theta) : (X_1,R_1,K_1) \to (X_2,R_2,K_2),$$

if M_1 is K_1-torsion free then f_*M_1 is K_2-torsion free.

<u>PROOF</u>. Apply the commutativity of the foregoing diagram to M_1 and $M_2 = f_*M_1$. ■

Let $(f,\Theta) : (X_1,R_1) \to (X_2,R_2)$ be a morphism of (pre) ringed spaces; let f_*, resp. f^* be the associated direct, resp. inverse, image functor, then :

<u>LEMMA</u> IV.2. Suppose that $\Theta : R_2 \to f_*R_1$ is a ring morphism which makes f_*R_1 into a flat R_2-algebra and suppose that E is an injective R_1-Module. Then f_*E is an injective R_2-Module, where the

R_2-Module structure of f_*E is defined by Θ.

PROOF. It is well known that f^* is the left adjoint of f_*. Now, E being injective, $\mathrm{Hom}_{R_1}(-,E)$ is an exact functor in $\sigma(R_1)$ (resp. $\pi(R_1)$).
We have to show that $M_2 \to \mathrm{Hom}_{R_2}(M_2,f_*E)$ is an exact functor. But since :

$$\mathrm{Hom}_{R_2}(M_2,f_*E) \cong \mathrm{Hom}_{R_1}(f^*M_2,E),$$

and since f^* is exact by the assumption on Θ, the foregoing statement is obvious. ■

PROPOSITION IV.3. Let

$$(f,\Theta) : (X_1,R_1,K_1) \to (X_2,R_2,K_2)$$

be a morphism of (pre) kerneled spaces, let $M_\nu \in \sigma(R_\nu)$ (resp. $\pi(R_\nu)$), $\nu = 1,2$, and let

$$(f,\psi) : (X_1,M_1) \to (X_2,M_2)$$

be a morphism of Module structured spaces, semilinear with respect to (f,Θ). If Θ is flat then (f,ψ) may be extended to an essentially unique morphism :

$$(X_1,Q_{K_1}(M_1)) \to (X_2,Q_{K_2}(M_2))$$

making the following diagram commute :

$$\begin{array}{ccc} (X_1,M_1) & \xrightarrow{(f,\psi)} & (X_2,M_2) \\ \uparrow & & \uparrow \\ (X_1,Q_{K_1}(M_1)) & \longrightarrow & (X_2,Q_{K_2}(M_2)) \end{array}$$

<u>PROOF</u>. Let E_1 be an R_1-injective hull of M_1. By the foregoing lemma, f_*E_1 is injective; therefore we obtain a commutative diagram of R_2-morphisms :

$$\begin{array}{ccccc} \overline{0} & \longrightarrow & M_2 & \longrightarrow & E_2 \\ & & \downarrow \psi & \swarrow g & \\ & & f_*M_1 & & \\ & & \downarrow & & \\ & & f_*E_1 & & \end{array}$$

where E_2 is the R_2-injective hull of M_2. So we have :

$$\begin{array}{ccc} (X_1,M_1) & \xrightarrow{(f,\psi)} & (X_2,M_2) \\ \uparrow & & \uparrow \\ (X_1,E_1) & \xrightarrow{(f,g)} & (X_2,E_2) \end{array}$$

Since for $\nu \in \{1,2\}$:

$$Q_{K_\nu}(M_\nu) = M_\nu/K_\nu(M_\nu) \underset{E(M_\nu/K_\nu(M_\nu))/(M_\nu/K_\nu(M_\nu))}{\times} K_\nu(E(M_\nu/K_\nu(M_\nu)))$$

we may use Lemma IV.1. and the fact that quotients of (pre) structured spaces are well defined to obtain a morphism $\phi : Q_{K_2}(M_2) \to Q_{K_1}(M_1)$, i.e. a morphism :

$$(f,\phi) : (X_1,Q_{K_1}(M_1)) \to (X_2,Q_{K_2}(M_2)),$$

which is easily seen to be unique up to isomorphism, making the desired diagram commute. ■

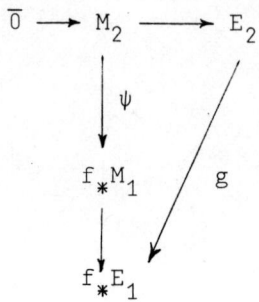

COROLLARY. The above proposition describes the functorial behaviour of the localization functor under change of base space by morphisms of (pre) kerneled spaces. Since R is always R-flat, functoriality of $Q_K(-)$ in $\sigma(R)$ $(\pi(R))$ follows from the foregoing result.

IV.2. Affine Schemes

Let us first recall some results and definitions from [2] and [23].

Assume A is a ring with unit and M a two-sided A-module. Its center is defined by

$$Z_A(M) = \{m \in M, \forall a \in A, am = ma\}.$$

The module M is said to be an A-bimodule in the sense of Artin or simply an A-bimodule, when no ambiguity arises, if M is generated as an A-module by $Z_A(M)$. If A is commutative, then an A-bimodule is simply a module in the usual sense. A morphism $\phi : M \to N$ of A-bimodules is by definition a homomorphism of two-sided A-modules. Of course $\phi(Z_A(M)) \subseteq Z_A(N)$. Conversely, if ϕ is a left (or right) A-linear homomorphism for which $\phi(Z_A(M)) \subseteq Z_A(N)$, then ϕ is a morphism of A-bimodules. We thus get a category $\underline{\underline{b}}(A)$, with objects the A-bimodules and with morphisms the morphisms of A-bimodules. It is additive, but in general non-abelian.

Let $f : A \to B$ be a ring-morphism. Following Procesi [23] we call f an extension (resp. central extension) or we say that B is an A-algebra (resp. central A-algebra) if B is generated as an A-module by $Z_A(B) = \{b \in B; \forall a \in A\ f(a)b = bf(a)\}$ (resp. $Z(B) = Z_B(B)$). Hence B is an A-algebra if it is an A-bimodule. The main example of an A-algebra (resp. central A-algebra) is the algebra $A\{X_i\}$ (resp. $A[X_i]$) of noncommutative (resp. commutative) polynomials coefficients in A. Moreover, one easily sees that each A-algebra (central A-algebra) is a quotient of some $A\{X_i\}$ (resp. $A[X_i]$). It is evident that one thus defines categories $\underline{Alg}_A$ (resp. $\underline{Alg}_A^c$) having for objects A-algebras (resp. central A-algebras) and extensions for morphisms. Note that we can always assume that the extensions considered are injective :

LEMMA IV.2.1. Let $f : A \to B$ be a ring morphism. The following are equivalent

(a) f is an extension;

(b) the canonical injection $i : f(A) \to B$ is an extension.

PROOF. Obvious. ■

Although, making use of the objects in $\underline{Alg}_A$ and $\underline{Alg}_A^c$ it is now possible to introduce relative schemes, we will restrict for simplicity's sake to the absolute case, leaving the relativisation as an easy exercise to the reader (cf. [36]).

Recall from Chapter I that the spectrum of an unitary ring A is by definition the set X = Spec(A) of all prime ideals of A. To an ideal I of A we associate the set $V(I) = \{P \in X, P \supset I\}$ which depends only on the radical rad I of I. The complement of V(I) in X is denoted by X_I, and one easily checks that the sets X_I are the open sets of the Zariski-topology on X. We will always assume X to be endowed with this topology.

PROPOSITION IV.2.2. (Procesi). Let $\phi : A \to B$ be an extension, then

(a) if $P \in \mathrm{Spec}(B)$, then $\phi^{-1}(P) \in \mathrm{Spec}(A)$;

(b) The induced morphism

$$^a\phi : \mathrm{Spec}(B) \to \mathrm{Spec}(A)$$

is continuous. ■

One thus obtains a contravariant functor spec from $\mathcal{P}$, the category of rings and extensions, to *Top* the category of topological spaces and continuous maps.

Assume from now on that A is left Noetherian. In I.3. it was explained how we can put on spec A a presheaf of rings $\underline{Q}_A^0$ by sticking to X_I the ring $\underline{Q}_A^0(X_I) = Q_I(A)$. We thus obtained a preringed space $\text{Spec}^0 A = (\text{spec}(A), \underline{Q}_A^0)$.

More generally, we can define for any $M \in$ A-mod a presheaf of left $\underline{Q}_A^0$-modules on spec(A) by sticking to X_I the module of quotients $Q_I(M)$ of M at $\kappa_{(I)}$, as follows immediately from the following lemma :

<u>LEMMA</u> IV.2.3. The set $\{\kappa_{(I)}$, I two-sided ideal of A$\}$ defines a local kernel functor on $\pi(A)$.

<u>PROOF</u>. Obvious. ■

Although except for the prime case this local kernel functor is not necessarily $\sigma(A)$-compatible, sheafification techniques yield a covariant functor

$$\underline{\underline{a}}Q_A^0(-) =: \underline{Q}_A(-) : \text{A-mod} \to \sigma(\underline{Q}_A)$$
$$M \to \underline{Q}_A(M)$$

where $\underline{Q}_A = \underline{\underline{a}}(\underline{Q}_A^0)$. This functor is exact if spec(A) has a basis of geometric t-sets.

<u>LEMMA</u> IV.2.4. For any left A-module we have

$$\Gamma(\text{spec}(A), \underline{Q}_A(M)) \cong M.$$

PROOF. As $\underline{Q}_A^0(M)$ is separated, we clearly have $\underline{Q}_A^0(M) \hookrightarrow \underline{Q}_A(M)$, hence it suffices to check that $\Gamma(\mathrm{spec}(A),\underline{Q}_A^0(M)) \cong M$, but this is obvious by the definitions of $\underline{Q}_A^0(-)$. ■

Spec defines a contravariant functor

$$\mathrm{Spec} : \underline{P} \to \underline{K}^s,$$

we will first need a series of lemmas on prime ideals.

LEMMA IV.2.5. For any prime ideal P of A we have $\kappa_{A-P}A \subset P$.

PROOF. This is obvious, as an element $x \in A$ is contained in $\kappa_{A-P}A$ iff there is s in A-P such that $sAx = 0$, but then $AsA.AxA = 0 \subseteqq P$, hence $AxA \subseteqq P$, i.e. $x \in P$. ■

LEMMA IV.2.6. Let P be a prime ideal of A, and let $j_{P,A} = j_P : A \to Q_{A-P}(A)$ be the canonical morphism, then

$$j_P^{-1}(Q_{A-P}(P)) = P.$$

PROOF. First, let us denote by $\delta(P)$ the set $\{x \in A, Kx \subseteq P$ for some left ideal K in $L(A-P)\}$. If $x \in j_P^{-1}(Q_{A-P}(P))$, then $j_P(x) \in Q_{A-P}(A)j_P(P) \cap j_P(A)$, i.e. there exist $q_i \in Q_{A-P}(A)$ and $p_i \in j_P(P)$ such that $j_P(x) = \sum q_i p_i$. Let us choose $K \in L(A-P)$ such that $Kq_i \subseteq j_P(A)$ and note that $Kj_P(x) \subseteq j_P(P)$, as $Kj_P(x) \subseteq \sum' Kq_i p_i \subseteq \sum' j_P(A)p_i \subseteq j_P(P)$; furthermore $Kx \subseteq P + \kappa_{A-P}A = P$, hence $x \in \delta(P)$, which proves that $j_P^{-1}(Q_{A-P}(P)) \subseteq \delta(P)$. Let us now show that $\delta(P) = P$. The inclusion $\delta(P) \supseteq P$ being obvious, let us prove the converse; take $x \in \delta(P)$, then there is $K \in L(A-P)$ such that $Kx \subseteq P$ and we can choose $K = (s)$ for some $s \in A-P$, so $AsA.AxA \subseteq P$ implying $x \in P$. Finally, that $P \subseteq j_P^{-1}(Q_{A-P}(P))$ follows from the commutativity of the following diagram :

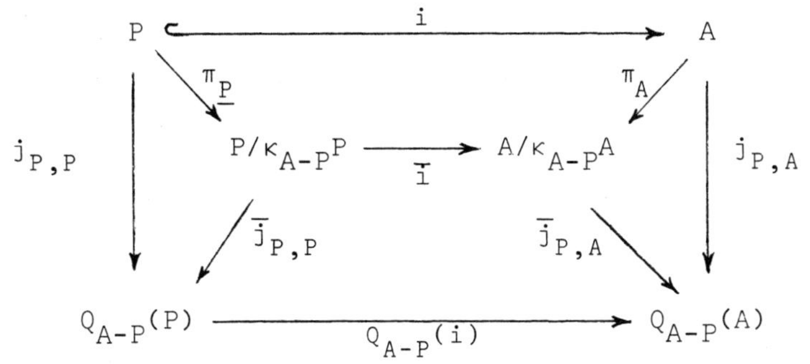

where i is the canonical inclusion. The induced map $\bar{i}$ is injective, as $\kappa_{A-P}P = \kappa_{A-P}A$ by the foregoing lemma, and $Q_{A-P}(i)$ is too. So, for $x \in P$, we have

$$j_P(x) = (j_{P,A} \circ i)(x) = Q_{A-P}(i) = j_{P,P}(x),$$

hence $j_P(x) \in Q_{A-P}(P)$. ∎

LEMMA IV.2.7. Let $\phi : A \to B$ be an extension and I an ideal of A. For any subset M of B let us denote by (M) the ideal of B generated by M. Then, for any $n \in \mathbb{N}$, we have $(\phi(I))^n = (\phi(I)^n)$.

PROOF. It suffices to prove this for $n = 2$, and in this case we have

$$\begin{aligned}(\phi(I))^2 &= B\phi(I)B\phi(I)B \\ &= B\phi(I)\phi(A)Z_A(B)\phi(I)B \\ &= B\phi(IA)\phi(I)Z_A(B)B \\ &= B\phi(I)^2B = (\phi(I)^2). \quad ∎\end{aligned}$$

COROLLARY IV.2.8. Let $\phi : A \to B$ be an extension, N a left A-module, M a left B-module and $\psi : N \to M$ a module-morphism, semi-linear with respect to ϕ. We obtain the following inclusion

$$\phi(\kappa_{(I)}N) \subseteq \kappa_{(J)}M$$

where $J = (\phi(I))$.

<u>PROOF</u>. If $n \in \kappa_{(I)}N$, there is $p \in \mathbb{N}$ s.t. $I^P n = 0$, hence

$$0 = \psi(I^P n) = \phi(I^P)\psi(n) = \phi(I)^P\psi(n)$$

so $(\phi(I)^P)\psi(n) = 0$, as

$$\begin{aligned} B\phi(I^P)B\psi(n) &= B\phi(I^P)\phi(A)Z_A(B)\psi(n) \\ &= B\phi(I^P)Z_A(B)\psi(n) \\ &= BZ_A(B)\phi(I^P)\psi(n) = 0. \end{aligned}$$

Hence, the foregoing lemma yields $J^P\psi(n) = (\phi(I))^P\psi(n) = (\phi(I^P))\psi(n) = 0$, i.e. $\underline{\underline{\psi}}(n) \in \kappa_{(J)}M$. ■

Let κ be a kernel functor in A-mod and $\phi : A \to B$ an extension; κ induces a kernel functor $\phi_*\kappa$ in B-mod, by defining for $M \in$ B-mod :

$$(\phi_*\kappa)(M) = \kappa(M_A)$$

where M_A is the A-module obtained from M by restriction of scalars. Clearly the idempotent filter of $\phi_*\kappa$ is given by

$$L(\phi_*\kappa) = \{I \text{ left ideal of } B, \phi^{-1}(I) \in L(\kappa)\}.$$

It is easy to prove that if ϕ is central or κ is a t-functor, then

$$Q_\kappa(\phi) = Q_\kappa(A) \to Q_\kappa(B)$$

is an extension (central !), cf. [6] where one first has to show that $[Q_{\phi_*\kappa}(B)]_A = Q_\kappa(B_A)$.
If one applies this to an extension $\phi : A \to B$, and $\kappa = \kappa_{(I)}$ then

if ϕ is central or X_I is a t-set, we get an extension $\phi_I : Q_I(A) \to Q_I(B)$.

<u>In the rest of this paragraph we will always assume that we are in one of these cases</u> : ϕ <u>is central or</u> spec(A) <u>has a t-basis</u>.

<u>LEMMA</u> IV.2.9. The functors $\phi_*\kappa_{(I)}$ and $\kappa_{(J)}$ coincide.

<u>PROOF</u>. Take $M \in$ B-mod, then $m \in \phi_*\kappa_{(I)}M$ iff there is $K \in L(\phi_*\kappa_{(I)})$ such that $Km = 0$. Let us prove that $K \in L(\kappa_{(J)})$. If $K \in L(\phi_*\kappa_{(I)})$, then there is $n \in \mathbb{N}$ s.t. $\phi^{-1}(K) \supseteq I^n$, but then $K \supseteq \phi(\phi^{-1}(K)) \supseteq \phi(I^n) = \phi(I)^n$, so $K \supseteq (\phi(I)^n)$, and lemma IV.2.7. yields $K \supseteq J^n$, i.e. $K \in L(\kappa_{(J)})$.
Conversely, if $K \in L(\kappa_{(J)})$ there exists n s.t. $K \supseteq J^n$; furthermore we may assume that K is two-sided, as $\kappa_{(J)}$ is symmetric; hence $\phi^{-1}(K) \supseteq \phi^{-1}(J^n) \supseteq \phi^{-1}(J)^n \supseteq I^n$. Finally, $I^n \subseteq \phi^{-1}(K)$ means that $\phi^{-1}(K) \in L(\kappa_{(I)})$ i.e. $K \in L(\phi_*\kappa_{(I)})$. ■

<u>COROLLARY</u> IV.2.10. With the notations of IV.2.8. we derive an isomorphism

$$\gamma_B : Q_I(B) =: Q_{\phi_*\kappa_{(I)}}(B) \overset{\sim}{\to} Q_J(B).$$

<u>THEOREM</u> IV.2.11. <u>Each extension</u> $\phi : A \to B$ <u>defines a morphism of ringed spaces</u>

$$\Phi = ({}^a\phi, \theta) : \mathrm{Spec}(B) \to \mathrm{Spec}(A).$$

<u>PROOF</u>. The continuous map ${}^a\phi$: spec(B) → spec(A) has been defined above, it is given by ${}^a\phi(P) = \phi^{-1}(P)$, for any $P \in$ spec(B); it is continuous as $({}^a\phi)^{-1}(X_I) = Y_J$ (X = spec(A), Y = spec(B)) of $J = (\phi(I))$. To construct the morphism $\theta : \underline{Q}_A \to ({}^a\phi)_*(\underline{Q}_B)$ it suffices to define it for the presheaves $\underline{Q}_A^0$ and $({}^a\phi)_*(\underline{Q}_B^0)$ and even

for a base of the topology on spec(A). So, if X_I is open in spec(A), we have $\Gamma(X_I, \underline{Q}_A^0) = Q_I(A)$ and $\Gamma(X_I, ({}^a\phi)_*(\underline{Q}_B^0)) = \Gamma(Y_J, \underline{Q}_B^0) = Q_J(B)$, hence we have to construct in a functorial way a morphism

$$\theta_I : Q_I(A) \to Q_J(B).$$

This morphism is the composition of

$$Q_I(A) \xrightarrow{\phi_I} Q_I(B) \xrightarrow{\gamma_B} Q_J(B)$$

(cf. IV.2.10). Verification of the functoriality in I is straightforward and left as an exercise. ■

For any prime ideal P of A we define $\Bbbk(P) = Q_{A-P}(A)/Q_{A-P}(P)$. A morphism (ϕ,θ) : Spec(A) → Spec(B) is <u>local</u> if for any $P \in$ spec(B), $Q = {}^a\phi(P) \in$ spec(A) the fibre morphism $\theta_P : Q_{A-Q}(A) \to Q_{B-P}(B)$ induces an injective morphism $\Bbbk(Q) \hookrightarrow \Bbbk(P)$. Let us note that in general $Q_{A-P}(A)$ is <u>not</u> a local ring; it is local e.g. if Q_{A-P} is a perfect localization, cf. next paragraph.

<u>THEOREM</u> IV.2.12. <u>The morphism</u> $({}^a\phi,\theta)$ <u>constructed in</u> IV.2.11. <u>is local</u>.

<u>PROOF</u>. A remark first : the morphism $\theta_P : Q_{A-Q}(A) \to Q_{B-P}(B)$ maps $Q_{A-Q}(Q)$ into $Q_{B-Q}(P)$ as $\phi_*\kappa_{A-Q} \leq \kappa_{B-P}$. This can be controled as follows. Let I be a two-sided ideal in $L(\phi_*\kappa_{A-Q})$, i.e. $\phi^{-1}(I) \in L(\kappa_{A-Q})$; as $\phi^{-1}(I)$ itself is also two-sided, there is $s \in A-Q$, such that $s \in \phi^{-1}(I)$, i.e. $\phi(s) \in I$. But this implies $I \in L(\kappa_{B-Q})$, for, if $\phi(s) \in P$, then $s \in \phi^{-1}(P) = Q$, which gives a contradiction. So a basis of $L(\phi_*\kappa_{A-Q})$ is contained in $L(\kappa_{B-Q})$, which proves our assertion. Consider the following commutative diagrams :

(1)

$$\begin{array}{ccc}
Q_{A-Q}(A) & \xrightarrow{\ \phi_P\ } & Q_{B-P}(B) \\
\downarrow{\scriptstyle p_Q} & & \downarrow{\scriptstyle p_P} \\
Q_{A-Q}(A)/Q_{A-Q}(Q) & \xrightarrow[\ \gamma\ ]{} & Q_{B-P}(B)/Q_{B-P}(P) \\
\| & & \| \\
\mathbb{k}(Q) & & \mathbb{k}(P)
\end{array}$$

(2)

$$\begin{array}{ccc}
A & \xrightarrow{\ \pi^Q\ } & A/Q \\
\downarrow & & \downarrow \\
Q_{A-Q}(A) & \xrightarrow[\ \pi_Q\ ]{} & Q_{A-Q}(A/Q) \\
& {\scriptstyle p_Q}\searrow & \uparrow{\scriptstyle \alpha} \\
& & Q_{A-Q}(A)/Q_{A-Q}(Q)
\end{array}
\qquad
\begin{array}{ccc}
B & \xrightarrow{\ \pi^P\ } & B/P \\
\downarrow & & \downarrow \\
Q_{B-P}(B) & \xrightarrow[\ \pi_P\ ]{} & Q_{B-P}(B/P) \\
& {\scriptstyle p_P}\searrow & \uparrow{\scriptstyle \beta} \\
& & Q_{B-P}(B)/Q_{B-P}(P)
\end{array}$$

where α and β exist by the left semi-exactness of Q;

(3)

$$\begin{array}{ccc}
Q_{A-Q}(A) & \xrightarrow{\ \phi_P\ } & Q_{B-P}(B) \\
\downarrow{\scriptstyle \pi_Q} & & \downarrow{\scriptstyle \pi_P} \\
Q_{A-Q}(A/Q) & \xrightarrow[\ \overline{\phi}_P\ ]{} & Q_{B-P}(B/P)
\end{array}$$

The morphism $\overline{\phi}_P$ is injective; indeed, it is the unique morphism which makes the following diagram commutative :

(4)

$$\begin{array}{ccccc}
A/Q & \xrightarrow{\ \overline{\phi}\ } & B/P & & \\
\downarrow{\scriptstyle j_{Q,A}} & & \downarrow{\scriptstyle j_{Q,B}} & \overset{j_P}{\searrow} & \\
Q_{A-Q}(A/Q) & \xrightarrow[\ Q_{A-Q}(\overline{\phi})\ ]{} & Q_{A-Q}(B/P) & \xrightarrow[\ \tau\ ]{} & Q_{B-P}(B/P
\end{array}$$

In this diagram $\overline{\phi}$ is injective, as $\phi^{-1}(P) = Q$, hence $Q_{A-Q}(\overline{\phi})$ is too; as B/P is κ_{B-P}-torsion free, cf. IV.2.5., $Q_{B-P}(B/P)$ is the κ_{B-P}-injective envelope of B/P, hence j_P is injective; as $\phi_* \kappa_{A-Q} \leqslant \kappa_{B-P}$ the A-module B/P is κ_{A-P}-torsion free, hence $j_{Q,B}$ is injective. From the essentiality of $Q_{B-P}(B/P)$ over B/P, we derive that τ is injective. Indeed, assume the converse, then τ has a kernel, say K, which differs from zero. But then $K \cap B/P \neq 0$, i.e. there is $b \neq 0$ in B/P s.t. $\tau \circ j_{Q,B}(b) = 0$, hence $j_P(b) = 0$ and $b = 0$, as j_P is injective. Contradiction.
Consider the following identities :

$$\beta \circ \gamma \circ p_Q \overset{(1)}{=} \beta \circ p_P \circ \phi_P \overset{(2)}{=} \pi_P \circ \phi_P \overset{(3)}{=} \overline{\phi}_P \circ \pi_Q \overset{(2)}{=} \overline{\phi}_P \circ \gamma \circ p_Q$$

where (-) refers to the diagram which provides us with the required identity; as p_Q is surjective, we conclude that $\beta \circ \gamma = \overline{\phi}_P \circ \alpha$, i.e. the following diagram is commutative

(5)
$$\begin{array}{ccc} \mathbb{k}(Q) & \xrightarrow{\ \gamma\ } & \mathbb{k}(P) \\ \alpha \downarrow & & \downarrow \beta \\ Q_{A-Q}(A/Q) & \xrightarrow[\overline{\phi}_P]{} & Q_{B-P}(B/P) \end{array}$$

But $\beta \circ \gamma = \overline{\phi}_P \circ \alpha$ is injective, as $\overline{\phi}_P$ and α are, so γ is injective.

■

We can prove a converse of this theorem.

<u>THEOREM</u> IV.2.13. <u>Let</u> Φ : Spec(A) → Spec(B) <u>be a local morphism such that the morphism of global sections</u> $\phi : B \to A$ <u>induced by</u> Φ <u>is an extension</u>, <u>then</u> $\Phi = ({}^a\phi, \theta)$, (notations as before).

<u>PROOF</u>. Assume $\Phi = (\psi, \theta)$: Spec(A) → Spec(B). For any $P \in$ spec(A) and $Q = \psi(P) \in$ spec(B), we get a stable morphism

$$\theta_P : Q_{B-Q}(B) \to Q_{A-P}(A)$$

and a morphism

$$\phi = \theta(\mathrm{Spec}(B)) : B \to A.$$

Consider the following commutative diagram, notations as in the foregoing theorem :

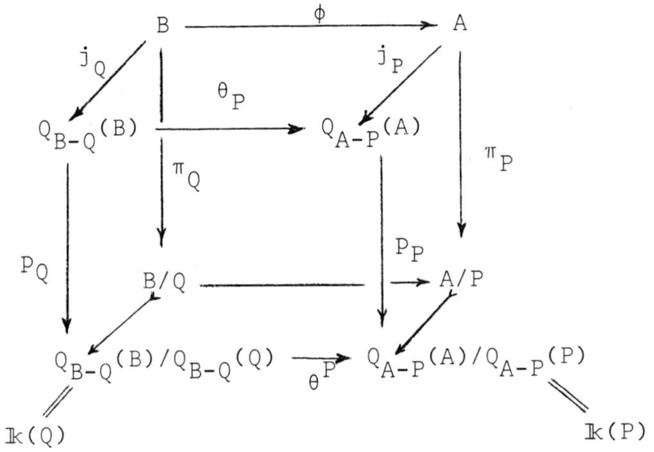

Lemma IV.2.6. yields that $A/P \to Q_{A-P}(A)/Q_{A-P}(P)$ and $B/P \to Q_{B-Q}(B)/Q_{B-Q}(Q)$ are injective morphisms. Furthermore, if we denote by $\pi_{P,P}$ (resp. $\pi_{Q,Q}$) the morphism

$$A \xrightarrow{\pi_P} A/P \longrightarrow \mathbb{k}(P)$$

$$(\text{resp. } B \xrightarrow{\pi_Q} B/Q \longrightarrow \mathbb{k}(Q))$$

then it is clear that $a \in P$ iff $\pi_P(a) = 0$, i.e. iff $\pi_{P,P}(a) = 0$ (resp. $a \in Q$ iff $\pi_{Q,Q}(a) = 0$). Let us prove that for any $P \in \mathrm{spec}(A)$, ${}^a\phi(P) = \psi(P)$, i.e. $\phi^{-1}(P) = Q$. This can be seen as follows : it is clear that

$$\theta^P\pi_{Q,Q} = \theta^P p_Q j_Q = j_P\theta_P j_Q = p_P j_P \phi = \pi_{P,P}\phi$$

and by hypothesis the map θ^P is injective, so

$$a \in \phi^{-1}(P) \Leftrightarrow \phi(a) \in P \Leftrightarrow \pi_{P,P}(\phi(a)) = 0 \Leftrightarrow \theta^P(\pi_{Q,Q}(a)) = 0$$

$$\Leftrightarrow \pi_{Q,Q}(a) = 0 \Leftrightarrow a \in Q.$$

Finally, if ϕ is central or spec(A) has a t-basis, ϕ induces a fibre map

$$\phi_P : Q_{B-Q}(B) \to Q_{A-P}(A)$$

and as there is a unique morphism $Q_{B-Q}(B) \to Q_{A-P}(A)$ which makes the following diagram commutative :

$$\begin{array}{ccc} B & \xrightarrow{\phi} & A \\ \downarrow{\scriptstyle j_Q} & & \downarrow{\scriptstyle j_P} \\ Q_{B-Q}(B) & \longrightarrow & Q_{A-P}(A) \end{array}$$

we easily see that for each $P \in$ spec(A) the morphisms ϕ_P and θ_P coincide. But then $\phi = \theta$, as follows from the next lemma. ■

LEMMA IV.2.7. Let $(\phi_1\theta_1)$ and $(\phi_1\theta_2)$ be two morphisms between the ringed spaces (X_1,R_1) and (X_2,R_2). These morphisms induce for each $x \in X$, $y = \phi(x)$ morphisms $\theta_{i,x}$, $i = 1,2$, between corresponding stalks

$$\theta_{i,x} : R_{2,y} \to R_{1,x}$$

Assume that for any x the morphisms $\theta_{1,x}$ and $\theta_{2,x}$ coincide, then $(\phi_1\theta_1) = (\phi_1\theta_2)$.

PROOF. Routine ! ■

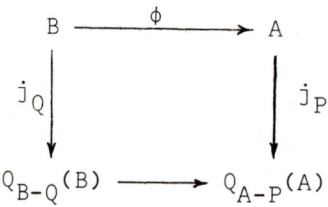

IV.3. Schemes

The results presented till now present two major shortcomings : in the first place there is no reason why $Q_{A-P}(P)$ should be an ideal of $Q_{A-P}(A)$, so in general $\mathbb{k}(P)$ will not even be a ring; on the other side the ring $Q_{A-P}(A)$ is not necessarily local, i.e. does not have a unique maximal ideal.
To remedy this defect and for technical reasons we will assume from now on that we work with geometric rings, i.e. rings for which spec(A) has a basis of geometric t-sets. We refer to [31] for the following result :

PROPOSITION IV.3.1. If A is a geometric ring then all stalks of the structure sheaf $\underline{Q}_A$ on spec(A) are geometric. In particular, if $P \in$ spec(A) then $Q_{A-P}(A)$ is a local ring with unique maximal ideal $Q_{A-P}(P)$.

Example : a Zariski central ring is geometric (cf. I.5.).

Let $\underline{\gamma}$ be the category with objects the geometrical rings and with morphisms the extensions between these rings; $\underline{G}$ will denote the category of geometric spaces, i.e. objects are locally ringed spaces and morphisms induce local extensions (in the usual sense) between the stalks.

THEOREM IV.3.2. *There is a contravariant functor*

$$\text{Spec} : \underline{\gamma} \to \underline{G}$$

assigning to each geometric ring A *the geometric space* Spec(A) = (spec(A), $\underline{Q}_A$).

PROOF. This follows immediately from the foregoing and Proposition IV.3.1. ■

A geometric space $(X,\underline{O}_X)$ is said to be an affine scheme if the following condition is satisfied : there exists a geometric ring A and an isomorphism

$$\Phi = (\phi,\theta) : \mathrm{Spec}(A) \overset{\approx}{\to} (X,\underline{O}_X)$$

in the sense that induces locally isomorphisms $\theta_x : \underline{O}_{X,x} \to Q_{A-\psi(x)}(A)$ which extend the global isomorphism $\Gamma(X,\underline{O}_X) \to A = \Gamma(\mathrm{spec}(A),\underline{Q}_A)$, and ψ is a homeomorphism of X on spec(A). A morphism of affine schemes is a morphism of geometric spaces which are affine.

A scheme is a geometric space $(X,\underline{O}_X)$ covered by open subsets U_α of X such that each induced geometric space $(U_\alpha,\underline{O}_X|_{U_\alpha})$ is affine. The full subcategory of $\underline{G}$ with objects the schemes is denoted by $\underline{Sch}$. If for $x \in X$

$$\theta_x : \underline{O}_{Y,\phi(x)} \to \underline{O}_{X,x}$$

is the stalk morphism induced by the scheme morphism $(\phi,\theta) : (X,\underline{O}_X) \to (Y,\underline{O}_X)$, then clearly θ_x induces an injective morphism $\mathbb{k}(\phi(x)) \to \mathbb{k}(x)$ where $\mathbb{k}(z) = \underline{O}_z/\underline{m}_z$, if $\underline{m}_z$ denotes the unique maximal ideal of the stalk $\underline{O}_z$. Note also that $\mathbb{k}(z)$ is simple but not necessarily Artinian.

The following theorem can be proved grosso modo as in the commutative case :

THEOREM IV.3.3. Let A be geometric ring and $X = (X,\underline{O}_X)$, then we get a bijection

$$\mathrm{Hom}_{\underline{Sch}}(X,\mathrm{Spec}(A)) \overset{\sim}{\to} \mathrm{Hom}_{\underline{P}}(A,\Gamma(X,\underline{O}_X))$$

(cf. IV.2.12 and IV.2.13). ■

Let us conclude this section with some remarks on closed subschemes.

<u>DEFINITION</u>. Let $\mathcal{X} = (X,\underline{O}_X)$ be a scheme. A <u>closed subscheme</u> of $\mathcal{X}$ is a couple $(\mathcal{Y},\pi)$, where $\mathcal{Y} = (Y,\underline{O}_Y)$ is a scheme such that Y is a closed subset of X with canonical inclusion $i : Y \hookrightarrow X$ and $\pi : \underline{O}_X \to i_*\underline{O}_Y$ a surjective sheaf-morphism. If π is determined by the context, then we will say that $\mathcal{Y} = (Y,\underline{O}_Y)$ is a closed subscheme of $\mathcal{X}$.

To a closed subscheme $\mathcal{Y}$ of $\mathcal{X}$ we associate a sheaf of ideals of $\underline{O}_X$ as follows : $\mathcal{J}(\mathcal{X},\mathcal{Y}) = \text{Ker}\ \pi$. It is clear that $\mathcal{J} = \mathcal{J}(\mathcal{X},\mathcal{Y})$ determines $\mathcal{Y}$ upto isomorphism. Conversely, each sheaf of ideals $\mathcal{J}$ of $\underline{O}_X$ does not necessarily determine a closed subscheme of $\mathcal{X}$ -we need some coherence conditions on $\mathcal{J}$!

<u>DEFINITION</u>. A scheme morphism $(\phi,\theta) : (X,\underline{O}_X) \to (Y,\underline{O}_Y)$ is a <u>closed immersion</u> if the following conditions are satisfied :

(a) ϕ is closed and injective on the underlying topological spaces;

(b) for every $y \in Y$ the induced morphism

$$\theta_y : \underline{O}_{X,\phi(y)} \to \underline{O}_{Y,y}$$

is surjective.

This means that (ϕ,θ) induces an isomorphism of $(Y,\underline{O}_Y)$ on a closed subscheme of $(X,\underline{O}_X)$. We can generalize to the noncommutative case the following result :

<u>PROPOSITION</u> IV.3.4. Let A be a geometric ring and I an ideal of A. The surjection $\pi : A \to A/I$ defines a closed immersion

$$(\theta,{}^a\pi) : \mathcal{Y} = \text{Spec}(A/I) \to \mathcal{X} = \text{Spec}(A),$$

and the sheaf of ideals $\mathcal{J}$ satisfies

(a) $\Gamma(X_I,\mathcal{J}) = Q_I(I) = IQ_I(A)$;

(b) $\mathcal{J}_x = I(\underline{Q}_A)_x$, i.e. $\forall P \in \text{spec}(A) : \mathcal{J}_P = I.Q_{A-P}(A) = Q_{A-P}(I)$.

PROOF. The map ${}^a\pi$ defines a bijection of spec(A/I) onto V(I), the set of all prime ideals containing I. More generally, if $J \supset I$ is an ideal of A and $\bar{J} = J/I$, then the map ${}^a\pi$ maps $V(\bar{J})$ onto V(J), i.e. ${}^a\pi$ is closed. That the θ_y are surjective can be seen as follows : θ_y is a localization of $\pi : A \to A/I$, if $y = \bar{P} \in$ spec(A/I), then

$$\underline{O}_{X,{}^a\phi(y)} = Q_{A-P}(A) \text{ and}$$

$$(\underline{O}_{Y,y})_A = (Q_{\bar{A}-\bar{P}}(\bar{A}))_A = Q_{A-P}(\bar{A}) = Q_{A-P}(A)/Q_{A-P}(I)$$

hence

$$\theta_y : Q_{A-P}(A) \to Q_{A-P}(A)/Q_{A-P}(I).$$

Finally we have

$$\begin{aligned}\Gamma(X_I,J) &= \mathrm{Ker}(\Gamma(X_I,\underline{Q}_A) \to \Gamma(X_I,\underline{Q}_{A/I}))\\ &= \mathrm{Ker}(Q_I(A) \to Q_{\pi(I)}(A/I))\\ &= \mathrm{Ker}(Q_I(A) \to Q_I(A/I)_A), \text{ as } Q_I(A/I) \hookrightarrow Q_{\pi(I)}(A/I)\\ &= Q_I(I) = IQ_I(A). \blacksquare\end{aligned}$$

THEOREM IV.3.5. (cf. [14]). Let X = Spec(A) and let Y be a closed subset of X. If J is the subsheaf of $\underline{Q}_A$ defining Y and $I = \Gamma(X,J)$, then $Y \cong$ Spec(A/I), i.e. we get a commutative diagram

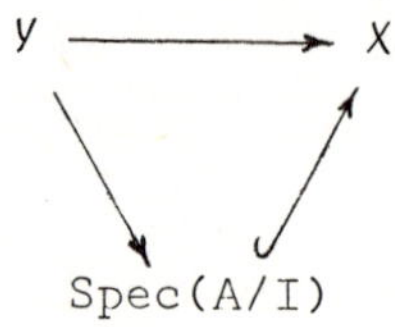

PROOF. Let $Y = (Y,\underline{O}_Y)$ and denote by (i,θ) the inclusion $Y \hookrightarrow X$. First note that i factorizes through spec(A/I), as by definition $j : \Gamma(\mathrm{spec}(A),\underline{Q}_A) = A \to \Gamma(Y,\underline{O}_Y)$ factorizes through $\bar{A}/I$, so we may assume that I = 0, i.e. the map j is injective. Now, as it is a closed subset of a (quasi-compact) affine scheme, the set Y is quasi-compact : there exists a finite number of open affines $\mathrm{spec}(A_i)$ covering Y. Let us show that i is a homeomorphism. As it is continuous, closed and injective, it suffices to prove its surjectivity. Since i(Y) is closed, there is an ideal J of A such that $i(Y) = V(J)$, so if we can show that $J \subset \mathrm{rad}(0)$, then $V(J) = V(0) = \mathrm{spec}(A)$ and i is surjective. Now, for $J \subset \mathrm{rad}(0)$ to hold, it suffices to show that each element of J is strongly nilpotent. Take $a \in J$, and consider a sequence $\{a_n\}_{n \in \mathbb{N}}$, where $a_{n+1} = a_n r_n a_n$, $r_n \in A$. We want to show that there is an index k such that for $n \geq k$ we have $a_n = 0$. For $y \in Y$ denote by ρ_y^Y the restrictions $\Gamma(Y,\underline{O}_Y) \to \underline{O}_{Y,y}$, then clearly $\rho_y^Y j(a_n) = 0$ for all $y \in V(J)$. Let us introduce the notation

$$\alpha_{n,i} = \rho_{A_i}^Y j(a_n),$$

then $\alpha_{n+1,i} = \alpha_{n,i}\rho_{A_i}^Y j(r_n)\alpha_{n,i}$. For each index i, and each $P \in \mathrm{spec}(A_i)$ we have $\rho_P^{A_i}(\alpha_n) = 0$, hence $\alpha_{n,i} \in \mathrm{rad}_{A_i}(0)$, so there is an index k with the property that for each i and each $n \geq k$ we have $\alpha_{n,i} = 0$. So for $n \geq k$, $j(a_n)$ has the following property : there exists a covering $\{\mathrm{spec}(A_i)\}$ of Y such that for all i :

$$\rho_{A_i}^Y(j(a_n)) = 0$$

But then $j(a_n) = 0$, as $\underline{O}_Y$ is a sheaf, and since j is injective we finally get $a_n = 0$ for $n \geq k$, hence a is strongly nilpotent.
It appears that we reduced the proof to the following situation : we have a single topological space Y and two sheaves $\underline{O}_X$, $\underline{O}_Y$ on it, together with a surjective map $\underline{Q}_A = \underline{O}_X \to \underline{O}_Y$. Let us prove the injectivity of this map. Take $y \in Y$, $j(y) = P \in \mathrm{spec}(A)$. We want to show that $\theta_y : \underline{O}_{X,P} = Q_{A-P}(A) \to \underline{O}_{Y,y}$ is injective. Assume the converse, then there is $a \in A$ in the kernel of θ_y. Consider again a finite, open covering

$\{spec(A_i) = X^{(i)}\}$ of Y, and extensions $\phi_i : A \to A_i$ such that the following diagram is commutative :

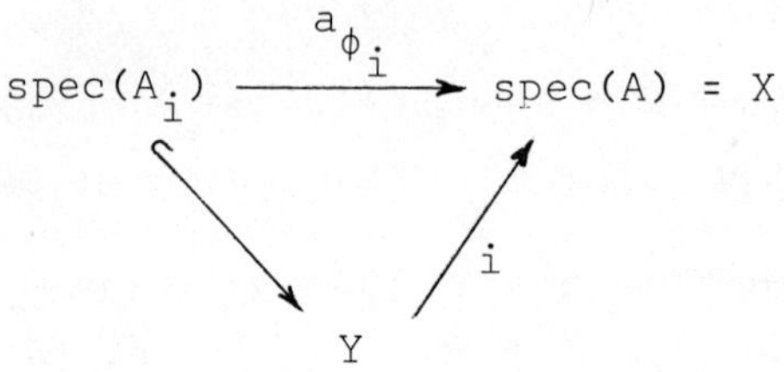

We know that for all $y \in Y$, $\rho^Y_y j(a) = 0$, hence there is an open set X_I, $I \not\subset P$ such that for all $y \in U = i^{-1}(X_I)$ we have $\rho^Y_y j(a) = 0$. Denote by $I^{(i)}$ the A_i-ideal $(\phi_i(I))$, then $U \cap X^{(i)} = X^{(i)}_{I^{(i)}}$ and $\Gamma(X^{(i)}_{I^{(i)}}, \underline{O}_Y) = Q_{I^{(i)}}(A_i)$. For every index i j(a) is zero in $Q_{I^{(i)}}(A_i)$. For each i choose $\alpha_i \in A_i$ giving j(a) on spec(A_i), then there exist integers x_i such that for each i we get

$$[I^{(i)}]^{n_i} \alpha_i = 0.$$

As the $X^{(i)}$ are finite in number there is a positive integer k such that for each i

$$[I^{(i)}]^k \alpha_i = 0.$$

But then, as in IV.2.8., it follows that for every index i we have $\rho^A_{A_i}(j(I^k a)) = 0$; as $\{spec(A_i)\}$ covers spec(A), this yields $j(I^k a) = 0$, and the injectivity of j finally gives $I^k a = 0$, hence $a \in \kappa_{A-P}A$, i.e. $j_P : A \to Q_{A-P}(A)$ maps a to zero, which proves the injectivity of θ_y. ■

Let us call a scheme $(X, \underline{O}_X)$ <u>reduced</u> if $\underline{O}_X$ has no strongly nilpotent sections, i.e. if $rad(\Gamma(U, \underline{O}_X)) = 0$ for all $U \in Open(X)$. One easily checks that it suffices to check this on an affine covering, and that the stalks of a reduced scheme have no strongly nilpotents. It is now an easy exercise to adapt the proof of the corresponding theorem in the commutative case, to show the following result :

PROPOSITION IV.3.6.

(a) Let $X = (X,\underline{O}_X)$ be a scheme and $Z \subseteq X$ a closed subset; among the closed subschemes $Z_\alpha = (Z,\underline{O}_X/J_\alpha)$ of X there is a unique reduced subscheme, say $Z_0 = (Z,\underline{O}_X/J_0)$;

(b) If $Z_1 = (Z,\underline{O}_X/J_1)$ is an arbitrary closed subscheme of X, then $J_1 \subseteq J_0$ and $J_0 = \underline{\text{rad}}(J_1)$ (i.e. $\Gamma(U,J_0) = \text{rad}\,\Gamma(U,J_1)$ for open, affine subsets.

If $Z = X$, we denote by X_{red} the unique closed reduced subscheme of X just constructed and thus define an idempotent endofunctor in *Sch* having the usual properties. This functor offers the possibility of introducing and studying varieties in noncommutative algebraic geometry, but an extensive study of this falls outside the scope of this text.

APPENDIX : ODDS AND ENDS

In this small chapter we aim to stimulate the reader's creative instinct, so we lay down the basis for a little theory, do some things with it and leave a lot to the reader.

A.1. A functor $p : \pi(R) \to \underline{\text{Ens}}$

Let R be a flabby Ring and consider the functor $p : \pi(R) \to \underline{\text{Ens}}$ defined by associating to $M \in \pi(R)$ the set of Points of M. Since pR obviously is a ring, $p(\pi(R)) = pR\text{-mod}$. If K is a kernel functor in $\pi(R)$ such that $L(K)$ is an idempotent filter then $pL(K)$ is an idempotent filter and as such it defines an idempotent kernel functor pK in pR-mod. It is straightforward to check that p is an exact functor; it is exactly this property that makes pK behave better in some ways, than the "Pointwise" functor PK of Section III.6.

Put $\kappa = pK$.

EXERCISE. Prove the following statements :

1. $pQ_{PK} = Q_{\kappa}p$ on $\pi(R)$
2. If K is a T(P)-functor in $\pi(R)$ then κ is a t-functor in pR-mod.

Although p is not an embedding of $\pi(R)$ into pR-mod, some information may be drawn for results modulo application of p in case one is interested exclusively in Points of presheaves and Pointwise localization.

A.2. Weakly Flabby Sheaves

Let X be a fixed topological space and let M be any presheaf over X. For each $x \in X$, define :

$(P^sM)_x = \{m \in M_x, m = \mu(x)$ for some global section $\mu \in M(X)\}$.

A presheaf M is said to be <u>weakly flabby</u> if, for all $x \in X$: $(P^sM)_x = M_x$. If M is a sheaf over X then the collection $\{(P^sM)_x, x \in X\}$ defines a subsheaf of M, which will be denoted by P^sM. In order to verify this statement it suffices to establish that for each $x \in X$ there is an open neighborhood U of x and a section $s \in M(U)$ such that $M_x^U(s) \in (P^sM)_x$, and that there is an open neighborhood V of x, $U \supset V$, such that for each $y \in V$ we have : $M_y^U(s) \in (P^sM)_y$. All this follows at once from the definition of the M_z^U, $z \in X$, and the fact that each element of the stalk at x is represented by a section over some open neighborhood of x. Hence, a sheaf is weakly flabby if and only if $M = P^sM$. It is clear that a flabby (pre) sheaf is weakly flabby.

<u>LEMMA</u> A.3.

1. Let M be a sheaf over X, then we have that $P^sM \cong \underline{\underline{a}}\ PM$.
2. If M is a weakly flabby (pre) sheaf and if N is a sub (pre) sheaf of M then M/N is weakly flabby.

<u>PROOF</u>. 1. Follows immediately from calculation of the stalks.
2. This statement amounts to the commutativity of the following exact diagram :

$$\begin{array}{ccccc}
M(X) & \longrightarrow & (M/N)(X) & & \\
\downarrow & & \downarrow & & \\
M_x & \longrightarrow & (M/N)_x = M_x/N_x & \rightarrow & 0 \\
\downarrow & & & & \\
0 & & & &
\end{array}$$

<u>LEMMA</u> A.4. Consider an exact sequence of sheaves over X :

$$\bar{0} \rightarrow M' \xrightarrow{f} M \xrightarrow{g} M'' \rightarrow \bar{0},$$

where M' is flabby and M" is weakly flabby, then M is a weakly flabby sheaf.

PROOF. For any $x \in X$ we have an exact sequence in Ab (or in R_x-mod, depending on the nature of the sheaves under consideration) :

$$0 \to M'_x \to M_x \to M''_x \to 0.$$

Furthermore, because of Proposition II.5.9. the following sequence is exact in Ab (R(X)-mod) :

$$0 \to M'(X) \to M(X) \to M''(X) \to 0.$$

So if $m_x \in M_x$ and $g_x(m_x) = m''_x$, then there is an $m'' \in M''(X)$ such that $M''^X_x(m'') = m''_x$. Pick an $m_1 \in g(X)^{-1}(m'')$, then $m_{1,x} - m_x \in f_x(M'_x)$. Since M' is a flabby sheaf it is certainly weakly flabby, hence we may choose a global section $m' \in M'(X)$ such that $f_x M'^X_x(m') = m_{1,x} - m_x$. Then $m = m_1 - f(X)(m') \in M(X)$ is a global section of M such that :

$$M^X_x(m_1 - f(X)(m')) = m_{1,x} - (m_{1,x} - m_x) = m_x,$$

hence M is weakly flabby. ■

Note : The above lemma holds for presheaves even if M' is only weakly flabby.

PROPOSITION A.5. Let K be a strictly local kernel functor in $\sigma(R)$, where R is a weakly flabby Ring, then the following statements are equivalent :

1. $L(K)$ is an idempotent filter
2. $L(K)$ has a filterbasis consisting of weakly flabby left Ideals.

PROOF. 2. ⇒ 1. Let I be a left Ideal of R such that $(I : \bar{\mu}) \in L(K)$ for every $\bar{\mu} \in J$, $J \in L(K)$. For $x \in X$, $(I_x : \mu_x) \supset (I : \bar{\mu})_x$ and since $(I : \bar{\mu}) \in L(K)$, with K being strictly local, we have that $(I : \bar{\mu})_x \in L(K_x)$, hence $(I_x : \mu_x) \in L(K_x)$. The latter holds for every $\mu_x \in J_x$ since J may be supposed to be weakly flabby by 2., moreover $J \in L(K)$ yields $J_x \in L(K_x)$ since K is strictly local. Hence $(I_x : \mu_x) \in L(K_x)$ for every $\mu_x \in J_x$, $J_x \in L(K_x)$; but this yields $I_x \in L(K_x)$ and $I \in L(K)$ follows by globalization.
1. ⇒ 2. Take a left Ideal $I \in L(K)$ and consider P^sI. If $\bar{\mu} \in I$, then $\underline{\underline{a}}$ $PI = P^sI$ yields that $\bar{\mu} \in P^sI$ and thus $(P^sI : \bar{\mu}) = R$, $R \in L(K)$ for all $\bar{\mu} \in I$, $I \in L(K)$. Then 1. implies $P^sI \in L'(K)$. ∎

REMARK. Let R be a coherent sheaf of left Noetherian rings and let K be a local $\sigma(R)$-compatible kernel functor in $\pi(R)$, then K^s is strictly local, moreover, the fact that $L(K^s)$ is an idempotent filter may now also be derived from Proposition III.3.3..

Let K be a kernel functor in $\sigma(R)$. A sheaf of R-modules M is said to be P^sK-free if $P^sK(M) = 0$, M is said to be P^sK-torsion if $P^sK(M) = M$. The latter implies that M is weakly flabby and K-torsion.
Note that P^sK is not left exact. Define P^sK-injectivity as in the Pointwise theory.

EXERCISE I. Prove the statements analoguous to Propositions III.6.3., III.6.4., III.6.5., III.6.6., III.6.8., for P^sK instead of PK.

EXERCISE II. Let R be a flabby sheaf of left Noetherian rings (hence coherent), let K be a local $\sigma(R)$-compatible kernel functor in $\pi(R)$; then K^s is strictly local and the Pointwise localization in $\pi(R)$ may (essentially by sheafification) be carried over to $\sigma(R)$

using P^sK for PK. One easily proves that $Q_{P^sK}(R)$ is a Ring and that for every $M \in \sigma(R)$, $P^s Q_{K^s}(M) \cong \underline{\underline{a}}\, Q_{PK}(PM)$ and this is a $Q_{P^sK}(R)$-Module.

The functor Q_{P^sK} is left semi exact and $T(P^s)$-functors may be investigated stalkwise.

REFERENCES

[1] ARTIN M., Grothendieck Topologies,
Seminar Notes, Harvard University, 1962.

[2] ARTIN M., On Azumaya algebras and finite dimensional representations of rings,
J. of Algebra, 11 (1969), pp. 532-563.

[3] BOURBAKI N., Algèbre I, Ch. 1 à 3.
Hermann, Paris, 1970.

[4] BOURBAKI N., Algèbre Commutative, Chap. I-VII,
Act. Sci et Ind. nos. 1290, 1293, 1308, 1314,
Hermann, Paris, 1961-1965.

[5] CONNELL I.G., A Natural Transform of the Spec Functor,
Journal of Algebra 10 (1968), p. 69-91.

[6] DELALE J-P., Sur le spectre associé à un anneau,
Sém. alg. non commutative, publ. Math. Orsay 44,
exp. n° 11.

[7] GABRIEL P., Des Categories Abéliennes,
Bull. Soc. Math. France 90 (1962), pp. 323-448.

[8] GODEMENT R., Théorie des Faisceaux,
Hermann, Paris, 1958.

[9] GOLAN J., Localization of Noncommutative Rings,
Marcel Dekker, inc. New York, 1975.

[10] GOLAN J., RAYNAUD J., VAN OYSTAEYEN F., Sheaves over the Spectra of Certain Noncommutative Rings.
Communications in Algebra,

[11] GOLDIE A.W., Localization in Non-Commutative Noetherian Rings,
J. of Algebra 5 (1967), pp. 89-105.

[12] GOLDMAN O., Rings and Modules of Quotients,
J. of Algebra 13 (1969), pp. 10-47.

[13] GROTHENDIECK A., Sur Quelques Points d'Algèbre Homologique,
Tohoku Math. J. (1958), pp. 119-221.

[14] GROTHENDIECK A., DIEUDONNE J.A., Eléments de Géometrique Algébrique I,
Springer-Verlag, Berlin-Heidelberg-New-York, 1971.

[15] HEINICKE A.G., On the Ring of Quotients at a Prime Ideal of a Right Noetherian Ring,
Canad. J. Math. 24 (1972), pp. 703-712.

[16] HERSTEIN I.N., Noncommutative Rings,
The Carus Math. Monographs 15, Math. Assoc. Amer. 1968.

[17] LAMBEK J., Lectures on Rings and Modules,
Waltham, Toronto, London, 1966.

[18] LAMBEK J., Torsion Theories, Additive Semantics and Rings of Quotients,
Lecture Notes in Math. 177 (1971).

[19] LAMBEK J., MICHLER G., The Torsion Theory at a Prime Ideal of a Right Noetherian Ring,
J. of Algebra 25 (1973), pp. 364-389.

[20] LESIEUR L., CROISOT R., Algèbre Noethérienne Non-commutative,
Mémor. Sci. Math. 154 (1963).

[21] MURDOCH D.C., VAN OYSTAEYEN F., A Note on Reductions of Modules and Kernel Functors,
Bull. Soc. Math. Belg. XXVI, 1974, pp. 271-281.

[22] MURDOCH D.C., VAN OYSTAEYEN F., Noncommutative Localization and Sheaves,
J. of Algebra 35 (1975), pp. 500-515.

[23] PROCESI C., Rings with polynomial identities,
Marcel Dekker inc., New-York 1973.

[24] RAYNAUD J., Localization et Topologies de Stone,
preprint 1975.

[25] SILVER L., Noncommutative Localization and Applications,
J. of Algebra 7 (1967), pp. 44-76.

[26] STENSTROM B., Rings of Quotients, An Introduction to Methods of Ring Theory,
Springer-Verlag, Berlin-Heidelberg-New-York, 1975.

[27] VAN OYSTAEYEN F., Compatibility of Kernel Functors and Localization Functors,
to appear, Bull. Soc. Math. Belgique.

[28] VAN OYSTAEYEN F., Localization of Fully Left Bounded Rings,
Comm. in Algebra 4 (3), 271-284 (1976).

[29] VAN OYSTAEYEN F., Pointwise Localization of Presheaves of Modules,
Indag. Math. 80 (1977), pp.

[30] VAN OYSTAEYEN F., Primes in Algebras over Fields,
J. Pure and Appl. Algebra 5 (1974), pp. 239-252.

[31] VAN OYSTAEYEN F., Prime Spectra in Noncommutative Algebra,
Lecture Notes in Math. 444 (1975), Springer-Verlag.

[32] VAN OYSTAEYEN F., VERSCHOREN A., Localization of Presheaves of Modules,
Indag. Math. 79 (1976), pp. 335-348.

[33] VERSCHOREN A., VAN OYSTAEYEN F., Localization of Sheaves of Modules,
Indag. Math. 79 (1976), pp. 470-481.

[34] VERSCHOREN A., A note on Strictly Local Kernel Functors,
to appear.

[35] VERSCHOREN A., Localization and the Gabriel-Popescu Embedding,
to appear.

[36] VERSCHOREN A., Les Extensions en Géométrie Algèbrique Non-commutative,
to appear.

SUBJECT INDEX